ONVIEW BOOKS
Published by Onview.net Ltd.
2016

Registered Office & Distribution
Onview.net Ltd. Frilford Mead. Kingston Road, Frilford.
Abingdon. Oxfordshire. OX13 5NX. England.

First Published in 2016 by Onview.net Ltd.

A CIP catalogue record for this book is available.

ISBN-13: 978-1539554752
ISBN-10: 1539554759

Micscape Magazine Yearbook 1

Micscape Magazine Selected Articles
from July 2014 to June 2016

To see all issues in full on the Internet use an Internet connection and enter the address below.

www.microscopy-uk.org.uk/mag/issueindex.html

or

www.micscape.org/mag/issueindex.html

CONTENTS

A "CONVENTIONAL" DIGITAL CAMERA FOR TAKING MICROSCOPY PICTURES AND/OR VIDEOS — ALEJANDRO ARIEL GARCIA ARRIAGA - *Mexico* - 6

MEASURING WITH A LASER BEAM — ALEJANDRO ARIEL GARCIA ARRIAGA 9

Review of: Universal Cell Phone Adapter Mount for Binocular, Monocular, Spotting Scope, Telescope, and Microscope — Mol Smith - *UK* - 13

Topical Tip: Using a stereo LED ringlight as a darkfield 'condenser' on a stereo microscope. — David Walker - *UK* - 15

Book review: 'Antoni van Leeuwenhoek: Master of the Minuscule' — David Walker - *UK* - 17

THE LITHOGRAPHY BUSINESS AT 54 HATTON GARDEN: FROM OBSCURITY TO CELEBRITY AND PROLIFIC SUCCESS — Peter B. Paisley - *Sydney, Australia* - 20

Reflections on studying *Spirogyra* - a classic school biology subject and plenty of interest for the hobbyist — David Walker - *UK* - 32

The riddle of the 'green streaks'. In search of the first microorganism which Antoni van Leeuwenhoek described — Wim van Egmond in collaboration with Frans Kouwets - *the Netherlands* - 38

The Sand Box — Kirsten Martin - *USA* - 43

The Diverse Ciliate Community — Jason Dinelli - *USA* - 47

Surfaces: Part 1 - Tools. — Richard L. Howey - *Wyoming, USA* - 52

Surfaces: Part 1 - Specimens — Richard L. Howey - *Wyoming, USA* 55

My Route to Microscopy and Crystal Pictures — Theo Wyatt - *UK* - 62

UNUSUAL MICROSCOPES: THE ELGEET ZOOM PROJECTION MICROSCOPE (Ingenious, Practical, and Extinct) — Manuel del Cerro and Dietmar R. Krause - *Princeton, NJ, USA* - 66

WINTER IS FOR MICROSCOPY III. SOME PROTOZOA — Anthony Thomas - *Canada* - 69

A simple differential stain of blood smears using black Quink® — Chris Thomas - *UK* - 72

WINTER IS FOR MICROSCOPY II MULTICELLULAR ALGAE — Anthony Thomas - *Canada* - 78

The Paramecium Enigma (Do one cell organisms have a form of intelligence?) — Mol Smith - *UK* - 81

Notes From The Editor (Country Contributions) — Editor 83

A close up view of two "Parrot Tulips". *Tulipa x hybride* (Abridged) Part 1 — Brian Johnston - *Canada* - 84

A CHEAP AND PRECISE SLICER FOR TEACHING BOTANY (and new adventures in my garden) — WALTER DIONI - *Durango (Dgo) Mexico* 90

TECHNICAL TIPS ON THE USE OF THE DOUBLE RAZOR BLADES SLICER — WALTER DIONI - *Durango (Dgo) Mexico* 95

The Dicotyledon Stem. For the beginner botanist from a beginner botanist — WALTER DIONI - *Durango (Dgo) Mexico* 99

The Strange And The Beautiful — Ian Walker - *UK* - 104

A Final Say — Mol Smith - *UK* - 112

Note: The way article titles are presented above is a reflection of each author's preference for the title of their article. Some prefer full capitals, others do not. We have preserved their preferences.

A Brief Introduction

Apart from the first few black and white photocopied duplications of a printed version of Micscape Magazine back in the late 1990s, this is the very first true physical representation of the online Ezine.

Micscape came to life in November of 1995. The Internet and the World Wide Web were in their infancy and two men, myself and David Walker, thought an opportunity existed in this evolving environment to engage other people in the hobby of Amateur Microscopy. It's now August 2016, and for twenty-one years or so, David Walker has unfailingly received articles and material from other like minded and unselfish people throughout the world, and published them on the Internet in the magazine.

The article library is vast and reaches across the entire spectrum of subjects studied with a microscope as well as diving into microscope manufacture, the history of microscopy, and different techniques.

A hardy bunch of core authors during various periods throughout those two decades have born the brunt of regular and profound article creation to attract the attention and work by less frequent authors. My role has been part author, but far more one of being the tech guy in the background facilitating the publication through my understanding of the Internet protocols and the technologies which drive it.

As a co-founder, I have become increasingly concerned by the way the Internet has evolved from a predominantly free information-access medium to one (at the present time) of commercial marketing and merchandising goods. The balance between the Internet's value as a publicly accessible world-wide library and its use as a marketing and demographic tool for placing advertising has tipped too far in one direction. Inevitably, money talks, and money and the making of it now spins the Internet into its own image.

I have also grown concerned about the rapid changes taking place to the technologies behind the Internet which drive future direction. The advent of mobile smart phones and vying operating systems along with competitive intent to win over a larger section of the populace to a particular marketing platform rather than a competitor's one, places the mid-to-near future of our Microscopy web site in danger of being made obsolete.

Many web sites which started with us back in 1995, or in that first decade, are no longer in existence on the Internet. Micscape is not a formal organisation, nor a company. When David and I step off the planet and into the abyss, the long term future of all the work created by so many people is in jeopardy of disappearing, despite plans in place to prevent that from happening—for at least a few more years after.

I believe: should all plans to prevent the ultimate disappearance of articles and content published in Micscape Magazine fail, we need to at least leave a physical trace (or traces) of our existence as a useful publication. To that end, and with permission granted by the authors within, this publication is the first such 'trace'.

It is an eclectic mix of articles published between July 2014 and June 2016. They may not be the best selection and the articles published here may not be the author's best work to represent their style and expertise. They are simply a set or articles I regarded as forming a good balance of novel interest to other enthusiast microscopists or the general public; a mix where I also found the easiest to re-format from a web based publication (manually) to a paper based one. Many significant authors' work is not included but they will be in future year books.

I see these publications more of a collector's bounty—an historic signpost to the fact we existed and 'we did this kind of thing'. And should the worst happen, and all the work (which still continues) disappear from the electronic space of the Internet, what you now hold in your hand, like a rare coin or relic from the past, may be the only glowing ember left of a very bright creative fire that hopefully warmed millions of people a little bit in the early 21st century.

All typos, errors, and mistakes herein will be my fault. I apologise to readers and the authors for all they discover. We simply do not have the resources to diligently transform web pages into high quality perfect printed publications. Time and money, as well as people to carry out such work is sadly missing.

I hope readers both now and in the future understand that the breadth and wealth of the microscopic world is mostly explored by non-professionals... people with a natural curiosity who are prepared to probe some of the finer engineering and secrets of nature. A worthy task by any set of values. I, for one, salute them all.

We will find time to publish more traces but for now, this is it. **Note:** *all material remains copyright of the original authors and is licensed non-exclusively here for publication.*

Mol Smith

A "CONVENTIONAL" DIGITAL CAMERA FOR TAKING MICROSCOPY PICTURES AND/OR VIDEOS

BY: ALEJANDRO ARIEL GARCIA ARRIAGA,
COACALCO DE BERRIOZBAL ESTADO DE MEXICO, MEXICO
Published June 2016

INTRODUCTION:

If we like microscopy, sometimes we may only possess just a microscope, possibly an old one, but nothing to record the observations. If a person has a small budget, things are worse. Though specialized microscope cameras are not too expensive and we can find one from 80 US dollars, on certain occasions the only thing we have is a digital camera for photography. We may wonder if it is possible to take pictures of our microscopy observations: the answer is, yes it is possible, and with a great quality, see below.

DEVELOPMENT:

I have a specialized microscope camera but I love to demonstrate that it is possible to do microscopy in many different ways using the conventional instruments we may have at home, and that microscopy is limited just to the imagination and wishes of the person making the studies.

Although I have seen some articles on adapting cameras to the microscope, I have never tried one until I purchased this Nikon Coolpix camera:

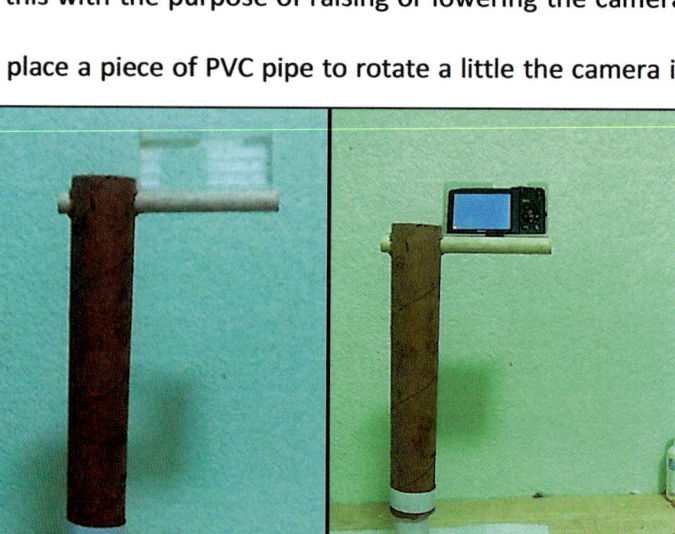

Since it is difficult to obtain a steady image just by holding it with the hands, I devised a holder which uses from the bottom to the top:

☐ a piece of wood,

☐ a length of the cardboard tube that typically supports aluminium foil or the film used to wrap foods; this with the purpose of raising or lowering the camera to adjust it to the eyepiece of the microscope,

☐ another tube with a hole at the top where I place a piece of PVC pipe to rotate a little the camera if necessary.

☐ and to hold the camera in place, a piece of those plastic soap boxes that are used in toilets, I created holes to pass the lens of the camera through to take the photo and another hole in the bottom. I place the camera in front of the microscope and zoom it to the point that it may be possible to see the image on the screen of the camera. I adjusted it by raising or lowering as needed.

I use it with both the 10x widefield (WF10x) eyepieces and with the 25 widefield (WF25x) eyepieces.

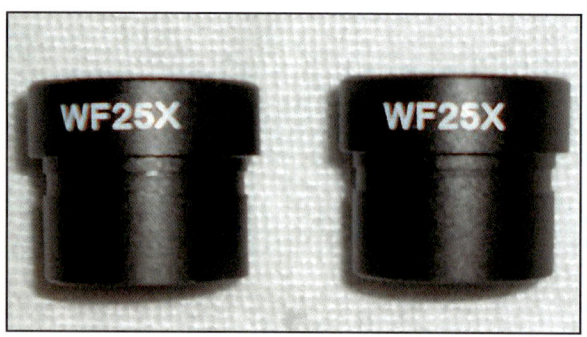

Top picture: the Coolpix Camera.
Picture above: my home made set up.
Left: the widefield eyepieces
Top right: the camera and attachment in place.

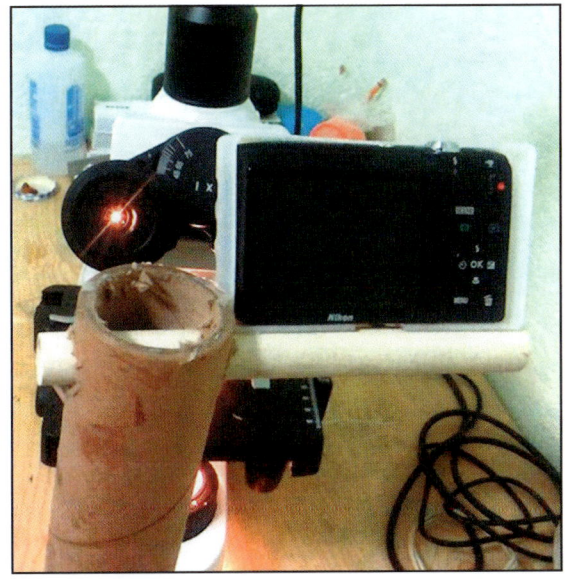

RESULTS:

Left: a portion of a spider leg 10x magnification plus WF10x in brightfield.
Below: a broken spider pouch and eggs 4x magnification plus WF10x in brightfield.

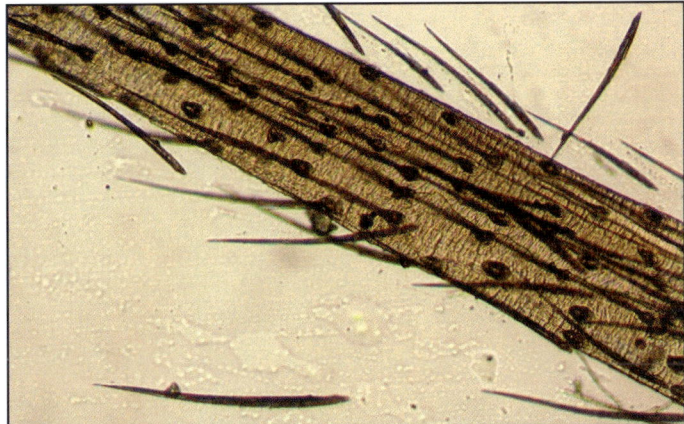

Below: onion stained with gentian violet with 4x magnification plus WF25X in brightfield.

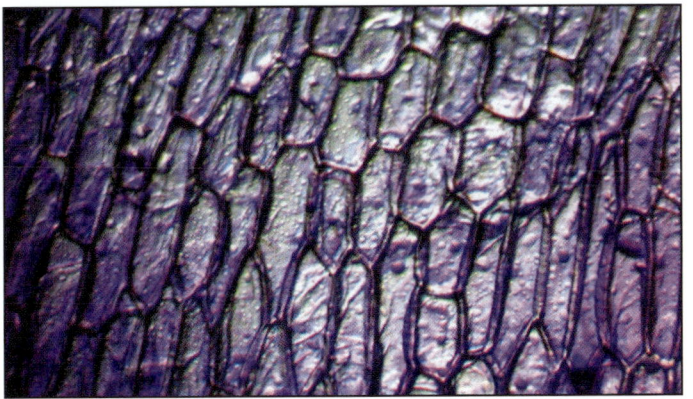

Daphnia pouch 10x magnification plus WF10X in brightfield.

It should be noted that magnifications quoted refer to the total magnification at the microscope . The process of introducing a digital camera and then post processing the images for both Internet and paper magazine issue render the magnification references meaningless to the pictures displayed here. {Editor}.

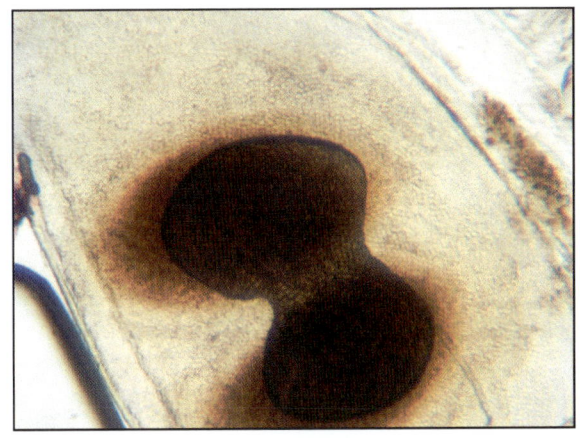

Below left: yellow lily pollen 10x plus WF10x eyepiece brightfield.
Below right: the previous spider's pouch but now 10x magnification plus WF25X in darkfield illumination.

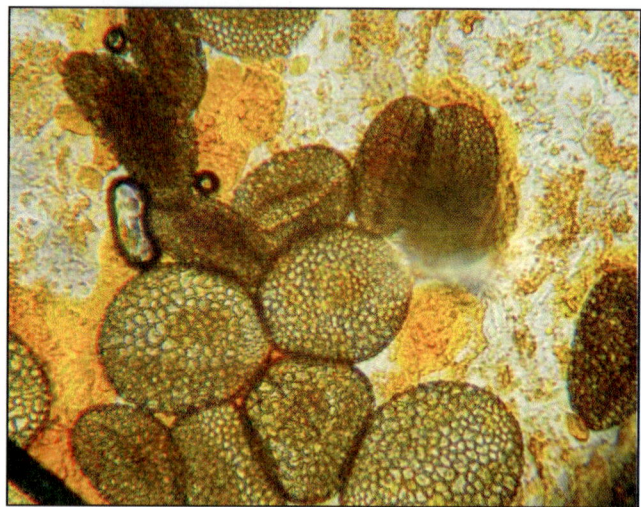 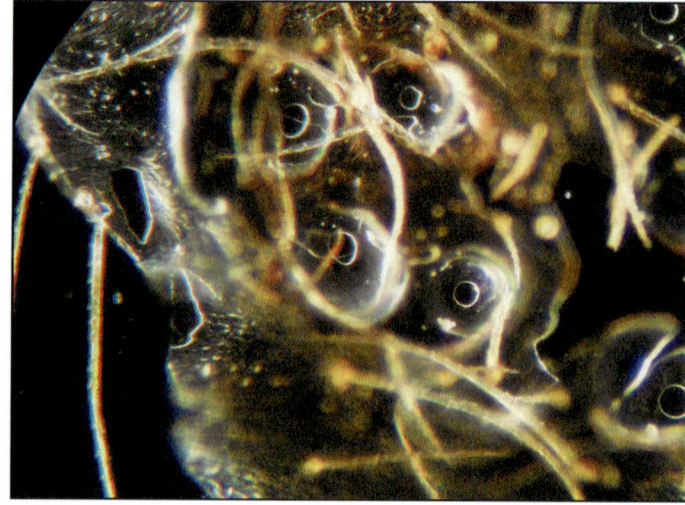

Below left: my mosquito 4X magnification plus WF10X in brightfield.
Below right: onion 10x plus WF25x in darkfield.

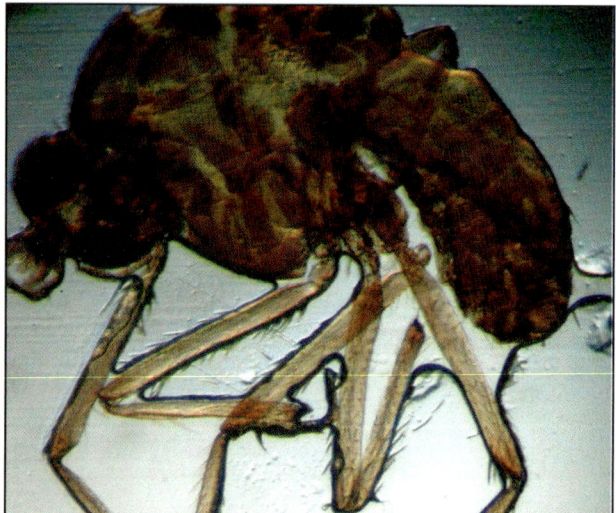 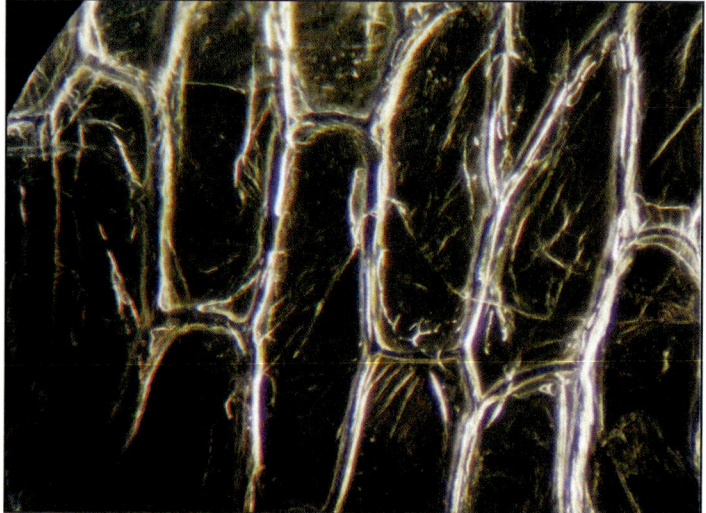

Below left: pollen of *Spathiphyllum wallisii* (Peace Lily) 10x plus WF10x in brightfield.
Below right: pollen of *Spathiphyllum wallisii* 40x magnification plus WF10x in brightfield.

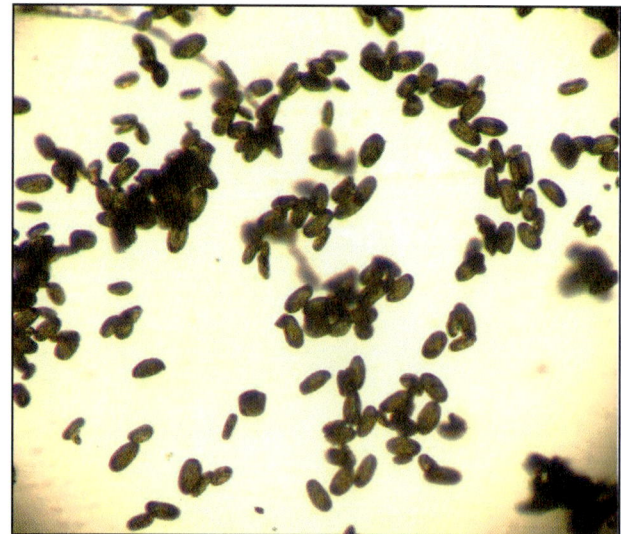 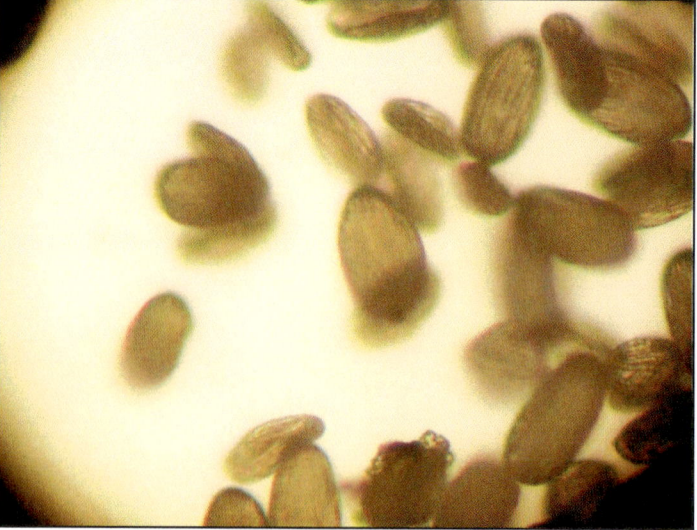

NOTE: I realized that the objects which are best captured by this camera are those that are well illuminated, so it is not good enough to catch pictures with epi-illumination. I tried this technique but it was not possible, at least with my system. I just got blurred images. I also tried a small piece of fern and I got no results from this because it was too "thick". That is a limitation. I present no videos because for the moment I had no samples showing movement.

CONCLUSION:
Maybe the quality of the pictures is not as good as those taken with a specialized camera but they are acceptable for recording our microscopy observations but obviously with the limitations expressed above.

Email author: doctor2408@yahoo.com.mx

MEASURING WITH A LASER BEAM
BY: ALEJANDRO ARIEL GARCÍA ARRIAGA
COACALCO DE BERRIOZÁBAL ESTADO DE MÉXICO, MÉXICO
Published June 2016

INTRODUCTION:
In my article about **Measurements of the micro world part 1. Calibrating the camera**, I used a micrometer slide that came with the installation software of my camera. After that I thought of presenting a second part to this article on the possibility of calibrating the camera without the help of a micrometer slide. This would be intended for those people that have a microscope camera but for one reason or another lack a micrometer slide. I did in fact succeed in calibrating my camera with the help of a conventional rule but just for the 4x and 10x objectives. The 40x and 100x were more difficult because of the thickness of the rule, nevertheless I did want to calibrate them.

So I was looking around for different subjects of a very small known size that can be used to calibrate the camera without the need of a micrometer. Finally, I thought of a hair and I said to myself "if I look for the average size of a hair, I would be able to study under the 40x, 100x lenses and would be able to calibrate too." But I found something much, much better. I entered into Google in Spanish the phrase "GROSOR DE UN CABELLO HUMANO" that means "the width of a human hair" and I was taken to many marvellous articles that explain how to measure with a good degree of accuracy any hair, using the laser beam of a conventional laser pointer. This opens up a world of possibilities because I could now have the object that I needed to calibrate my camera without the help of a micrometer and that I could measure by myself accurately.

Below: the web page I found.

Imágenes de grosor de un cabello humano   Notificar imágenes

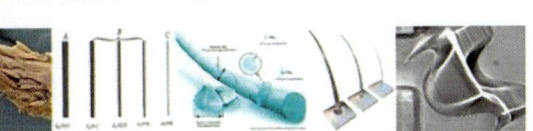

Más imágenes de grosor de un cabello humano

This way I learnt how to measure with a laser pointer and here is the explanation of the procedure.

El grosor del cabello | Secretos Expuestos
https://secretosexpuestos.wordpress.com/2013/06/29/4-el-grosor-del-cabello/ ▾
29 jun. 2013 - Cuando se habla del grosor del cabello se refiere al espesor de las hebras individuales de tu cabello en vez de cuanto cabello tienes sobre tu ...

Medición del diámetro de un cabello mediante difracción by Nacho ...
https://prezi.com/.../medicion-del-diametro-de-un-cabello-mediante-difraccion/ ▾
3 jun. 2015 - Principio de Babinet - d es el grosor del obstáculo, en nuestro caso el cabello - m es el orden de los máximos (en nuestro caso tomamos m=1)

Experimento casero: Cómo medir el grosor de un cabello
vicente1064.blogspot.com/2012/10/experimento-casero-como-medir-el-grosor.html ▾
1 oct. 2012 - El grosor de un cabello es del orden de magnitud de la longitud de onda de un láser ...

DEVELOPMENT:

To answer the question, how can we use a laser pointer as an instrument for measuring? The answer is very simple, the monochromatic light of a laser beam when it passes through a pair of holes (since light behaves as a wave) becomes diffracted - this is called double slit diffraction, and it can be seen at a distant point from where it is produced as showing an interference pattern of light waves. The bright spots are presented to each other at the same distance and are separated by dark spaces where the interference of the light becomes negative.

This beautiful and outstanding pattern is call Fraunhofer diffraction in honour of Joseph von Fraunhofer, although he was not exactly who developed the theory of this pattern, his equation is still used to explain the phenomenon. Another concept to consider within this phenomena is what is called the Babinet principle: this establishes that when a very thin opaque object is set in front of a monochromatic beam such as the one from the laser pointer, it is going to produce a pattern similar to that for the double slit mentioned above. The same will happen with a very thin slit. This phenomenon will allow us to obtain the size in terms of width of the object in front of the beam.

The equations to obtain a value are the following:

$$d = \frac{2*m*\lambda*D}{L} \quad \text{or} \quad d = \frac{m*\lambda*D}{L(1/2)}$$

d is the width of the object in front of the beam
m is the number of maximums one at each side of the centre, see below
λ is the wave length of the laser beam, it is said that for a green laser beam it is 532nm and for a red one it is 650nm
D is the distance from the object to the screen where it is projected
L is the distance between the maximums either side of the centre, see below.

Today we are going to consider the pattern produced by a very thin opaque object for example a hair, a thin wire, a hair of a pet fur, etc.

THE MATERIALS NEEDED:

- A tape measure
- A clear screen can be a wall, a cardboard, etc.
- A laser pointer, any colour
- A table and/or a holder to place the laser pointer
- Something to keep the laser pointer on could be some tape
- Obviously thin objects

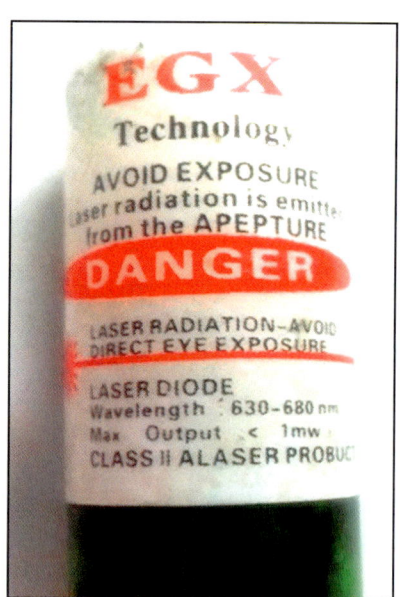

Note: regularly the wavelength of the laser beam is given by the manufacturer.

In my case since a wide range is given for the wavelength of the laser beam, I am going to take for granted the common data as an average of 650nm for the moment. But at the end of this paper I am going to demonstrate how to calculate more accurately the wavelength of my laser pointer and how an accurate value of the wavelength can give a better result.

RESULTS:

This is part of the pattern produced by a piece of hair when placed in front of the laser beam at 2 meters from the screen which in this case is a green piece of cardboard (*see top, next page*). With the help of a pen I marked the distance between the two maximums at each side of the central line and the result is as follows.

When we applied the formula mentioned :

$$d= \frac{m*\lambda*D}{L(1/2)}$$

First of all it is necessary convert every value into meters for an easy management of the calculation so, we obtain:

d= 1*0.000000650m*2m = 0.0000013m = 0.00011304347 = 0.1130 mm = 113 micrometers
　　0.023m /2　　　　　　0 .0115m

This is the width of the **wire strand employed.**

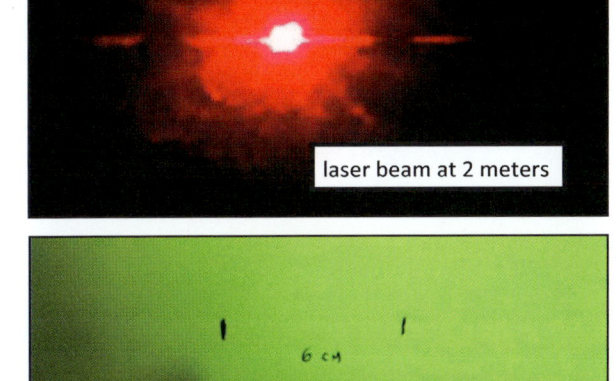

laser beam at 2 meters

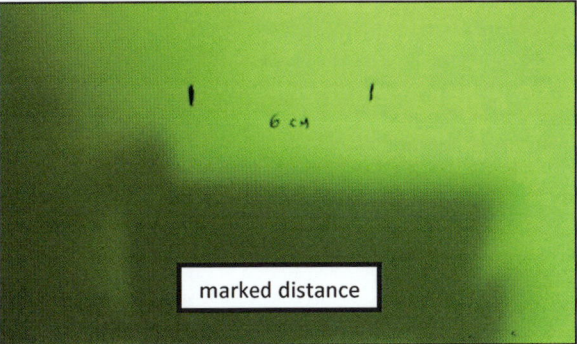

marked distance

Or **0.0433 mm, this is the width of the hair employed.**
The purpose of this is to get an accurate measure to be used without the need of a micrometer. To demonstrate the usefulness of this method I am going to place the same hair under the microscope and measure it with my already calibrated camera to show the accuracy of the method. Here is the result. As you can see it varies but is about 7 hundredths of a millimetre.

d= 1*0.000000650m*2m = 0.0000013m = 0.00011304347 = 0.1130 mm = 113 micrometers
　　0.023m /2　　　　　　0 .0115m

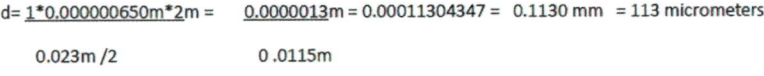

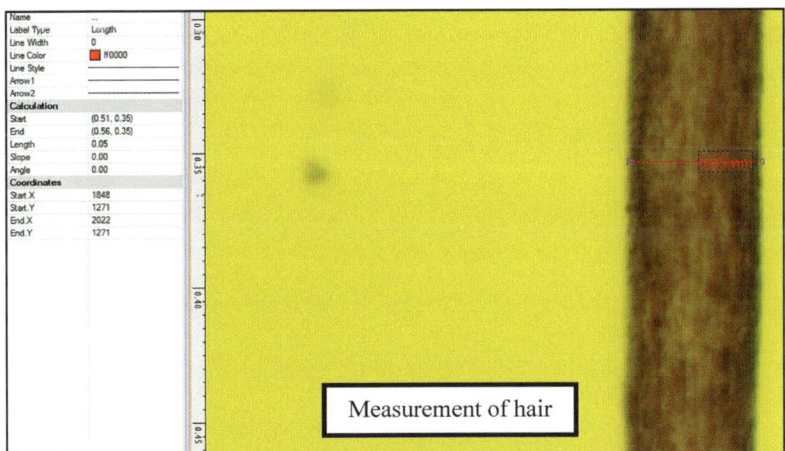

Measurement of hair

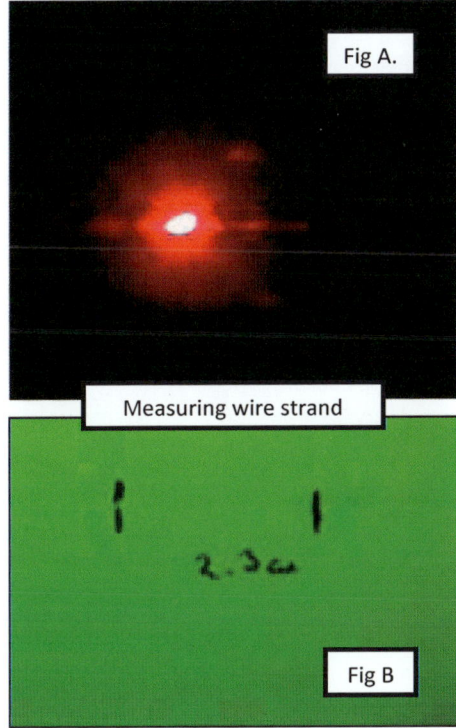

Fig A.

Measuring wire strand

Fig B

Let's try the next sample – a wire strand. (see fig A & B, right). This is the width of the **wire strand employed.** Here is the same wire strand with the microscope camera (*below*):

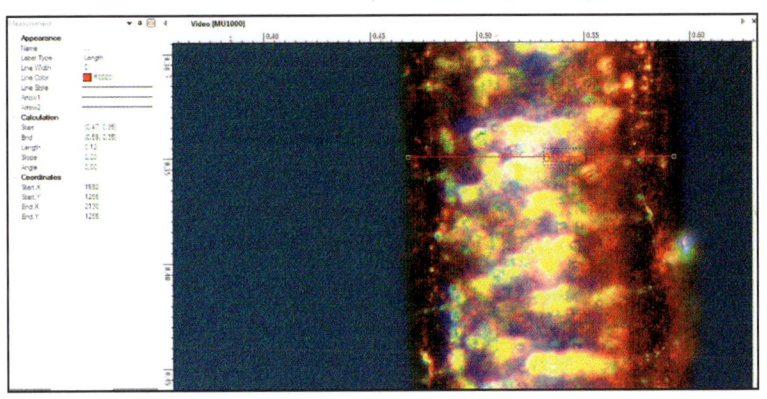

Also measured with the digital calliper.

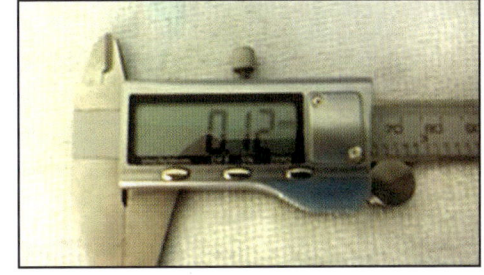

11

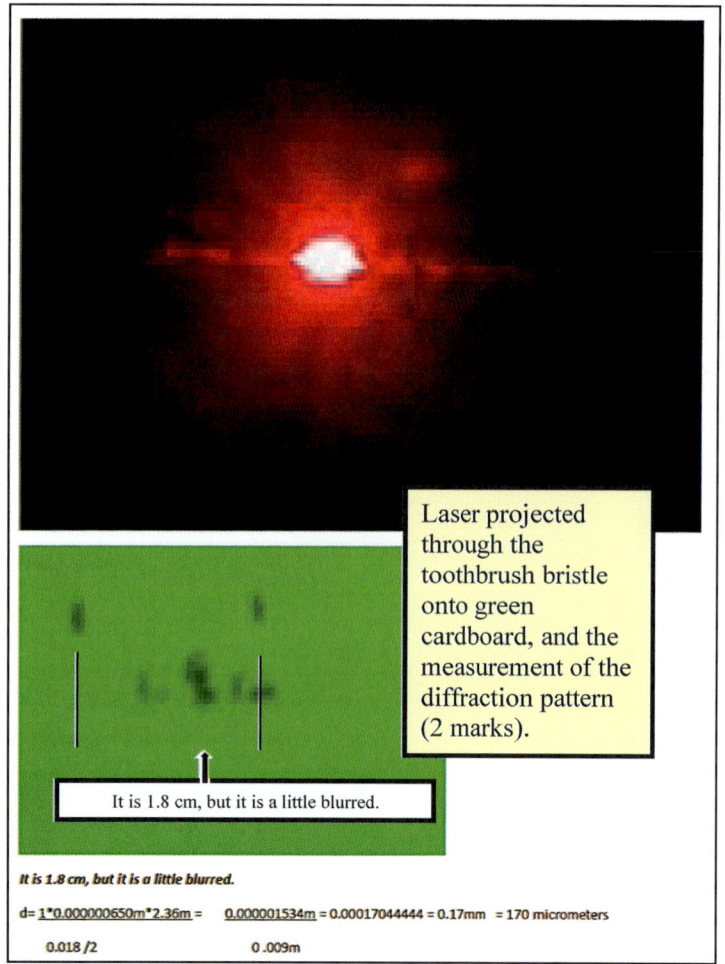

Laser projected through the toothbrush bristle onto green cardboard, and the measurement of the diffraction pattern (2 marks).

It is 1.8 cm, but it is a little blurred.

It is 1.8 cm, but it is a little blurred.

d= 1*0.000000650m*2.36m =   0.000001534m = 0.00017044444 = 0.17mm = 170 micrometers

0.018 /2   0 .009m

Finally the last sample, a toothbrush bristle, but now with a variation of the distance to the wall to see that no matter the distance the method is accurate (*see left*).

It is 1.8 cm, but it is a little blurred.

This is the width of the **toothbrush bristle employed**.

Now the bristle with the microscope camera.

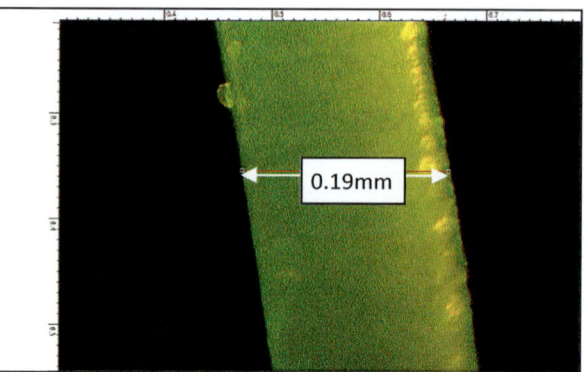

0.19mm

Also measured with the digital calliper.

Note: since I have never liked to take anything for granted, I am going to calculate the wavelength so that it may be closer to the real value. I count now with a pair of known values the wire and the bristle. So let's calculate the wavelength using both of them. So the formula above uses the wire width measured with the calliper:

$$d= m*\lambda *D \ = \lambda = L/2*d \ = \ 0.023m/2*0.00012m \ = 0.0115m*0.00012m \ = 0.00000069= 690nm$$

L(1/2)   m*D   1*2m

And with the bristle measured with the calliper:

$$= \ 0.018/2*.00018m \ = 0.009m*.00018m \ = 0.00000068644= 686.44nm \text{ very close to the last value of 690nm}$$

1*2.36m   2.36m

Finally when we correct the values for the samples to the wavelength found (690nm) the values are closer to the value given by the calliper and / or the microscope camera. Hair: 0.000046m or 0.046 mm or 46 micrometers, Wire: 0 .00012m or 0.12 mm or 120 micrometers, Bristle: 0.0001809 or 0.1809 or 180.9 micrometers.

CONCLUSION:
This is a good experiment to show a novel way to measure the microscopic world.
Email the author: *doctor2408@yahoo.com.mx*

REFERENCES:
http://www.cienciapopular.com/experimentos/medicion-del-grosor-del-pelo
https://www.youtube.com/watch?v=n1Y9MexS-Ok
https://www.youtube.com/watch?v=kpsN78mQ6YY
https://prezi.com/nxlzrkdtypsj/medicion-del-diametro-de-un-cabello-mediante-difraccion/

Review of: Universal Cell Phone Adapter Mount
for Binocular, Monocular, Spotting Scope, Telescope, and Microscope
by Mol Smith, UK
Published June 2016

With so many younger people (and many adults, too) now seeing their cell phones as indispensable items, I thought it worth looking at some of the good and not so good gadgets offered as attachments for exploring the small scale world.

From macro lens to fish eye, and miniature microscopes that attach directly to the smart phone, a range of inexpensive Chinese imports are for sale on Amazon. The first one which I thought might prove more useful than some of them was a Universal Cell Phone Adaptor for a range of optical viewing instruments, including a microscope. I won't say the brand name as I consider this not to be a manufacturer's product review but a generalised first look at the type of device. But this one (see above and right) cost £15.00 (UK pounds).

Most microscopists know it's probably best to use an SLR camera and attachment dedicated to that, or a dedicated microscope digital camera for serious work. Stability, vibration, and any micro movement of alignment is magnified tens to hundreds of times when working to capture video or still images. Precision and a rock-steady set-up is a first requisite!

However, it's likely that young people with busy, fast lives, are more likely to use a mobile camera and are probably quite satisfied with a less than perfect shot of something on the stage of a budget microscope, so I carried out the review with such an instrument (an entry-level monocular microscope).

The attachment looks fairly well made. Support for your mobile phone is by way of an adjustable clamp which has soft inner pads to protect the edges of the phone. The clamp attached to a bracket via a sliding screw in a slot, allowing the phone holder to move vertically and swivel on itself 360 degrees.

Attachment to the microscope is by means of a thumbscrew clamp which slides over the eyepiece to be gently closed to grip it. The pictures (right) demonstrates the build.

No instructions were included, in typical 'just try' mode, I clamped the thing onto the microscope and put the phone into the holder... thus (See right). I thought once I had it on the microscope, I would loosen the clamps to align the phone camera with the centre of the eyepiece lens. My phone is an LG G5. It has a brilliant set of cameras, but most smart phones, I believe, will work with this attachment. See right. The eyepiece lens 'rattles' a bit in the tube on this microscope, so I use a slip of paper wrapped 3/4 of a turn around the eyepiece to firm up the fit.

But that didn't work. Too much messing around was needed to try and align the phone both in a perpendicular plane and then to line up the tiny camera lens to the eyepiece centre. It's best to pre-align the eyepiece with the camera on the phone, by putting the phone in the clamp and the eyepiece in the thumbscrew clamp. Once you think you have it centred, pop it in the tube. You'll see me putting that slip of paper on again here. Not quite fully aligned but it will do for now to test the quality of the images. So, let's put a specimen slide on the stage: a Water Beetle. You can watch me fiddling around with the light, substage iris aperture and condenser, and the focus controls. The Moiré effect is the result of my SLR filming the camera screen and the rainbow patterning is not visible on the camera screen. I filmed this from the phone camera which you can watch on the next page to see the quality yourself from video stills.

Here we are then—direct footage taken with the cell phone. The original clip is HD 1920 x 1080. I've shrunk it here and degraded the quality slightly through re-encoding the video to a smaller size. And then sampling out stills and up-scaling them for this paper version of Micscape. I think as a 'quick and dirty' tool to engage youngsters, the quality is good enough, although I think any child say, below the age of ten years would struggle to get it properly aligned on the microscope. It takes a lot of patience.

My phone can record at a rapid rate of 114 frames per second so using it to record fast events on a micro-slide (pond life..? Where critters often move very rapidly) would prove an advantage as they can be played back in very slow but real slow motion. My camera also has time lapse functionality which would prove effective say, in recording something like a brine shrimp egg 'hatching'.

And not to boast about the superiority of the LG G5 over Apple and Galaxy phones, it has a mode where video can be shot using recording on both forward and backward cameras. This means a video is create where you can be seen narrating what you can see. Useful for teaching others.

Overall verdict. Yes! A useful tool if your smart phone, like many today, is out-performing expensive dedicated video and DSLR cameras. You already have the phone, you have the microscope, fifteen pounds for this will give further advantages to exploring the small-scale world.

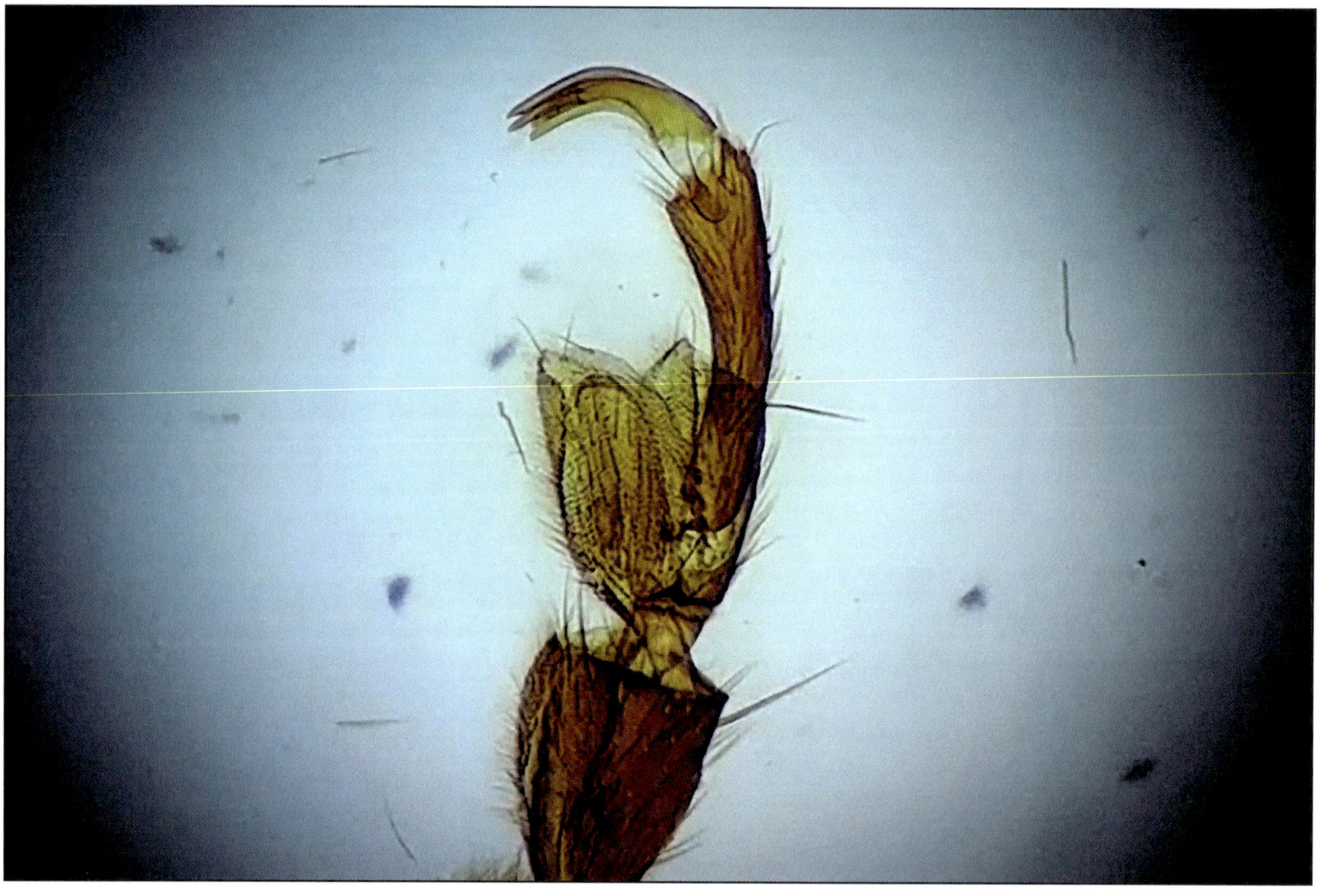

For paper version of Micscape only:

I'll probably use the adaptor quite a lot. It does allow for rapid social sharing and it's easy to simply connect the phone to my PC and transfer the files—both video and stills. Most of my needs are non-specialist and my intentions are more to get non-microscopists to become more aware of the world they live in which I firmly believe is built bottom up. By using the type of equipment the beginners and youngsters are likely to use, I feel I stand the best chance of making others aware of this insight. The main issue with the adaptor is the failure of being able to introduce another functional lens between the eyepiece and the camera of the phone. Without it, the phone's digital zoom is used, which we know is just empty magnification.

Topical Tip: Using a stereo LED ringlight as a darkfield 'condenser' on a stereo microscope.

by David Walker, UK

Ringlights are very useful for 'fit and forget' incident lighting on stereo microscopes. More recent models tend to use one or more rings of LEDs, although the models based on either fluorescent tubes or fibre optic light sources are also available. I use the YK-B144T model with 144 LEDs which was reviewed in the July 2012 issue of *Micscape*. The ten intensity settings and ability to control the four light quadrants independently allows useful modelling when desired.

I hadn't given much thought to the potential alternative uses of ringlights until read a message in the Yahoo 'Microscopes' Forum by Ricardo Tsukamoto where he highlighted and linked to a paper by K F Webb entitled 'Condenser-free contrast methods for transmitted light microscopy' (JRMS, 2015, vol. 257 (1), 8-22). The paper described the use of fabricated rings of LEDs for various compound microscopy techniques without the need of a condenser. Techniques illustrated included phase, darkfield and Rheinberg. The paper made me wonder how well the ringlight worked as a darkfield condenser on my stereo as I didn't currently have a ready way of doing this. With a little background reading, I wondered why I'd never tried this before as Brian Bracegirdle in his invaluable book *Scientific PhotoMACROgraphy* (1995) describes and illustrates the successful use of fluorescent and fibre optic ringlights for darkfield macroscopy studies.

So without further ado, I tried the 144 LED ringlight for darkfield on my Leica S8 stereo. The three concentric rings of LEDs are aligned to face inward to provide a bright central spot of light at the typical working distance range of a stereo of ca. 80 - 100 mm. The sample therefore ideally needs to be at least 80 mm from the ringlight. The setup I tried is shown below.

The setup gives an excellent velvety black darkfield for the full range of optical settings of the 1-8X zoom on the stereo. The LEDs are very bright and only set near its minimum intensity at lower mags. For photography it's best to set a level sufficient for good darkfield but without excess glare. Because the LEDs face inwards, I don't find any shrouding is required to reduce visual glare while at the stereo although card baffles could be readily added.

Above. Darkfield is very useful for spotting small organisms, either planktonic or epiphytic, for isolation to study under the compound. Also a very pleasing way to study the larger life in situ under less stressed conditions for the fauna than a cell under a compound microscope may give. LEDs are a cold light source as well.

Above. Darkfield setup on a stereo microscope. The control box of the 144 LED ringlight offers ten intensity settings and with each quadrant independently switchable. The LEDs face inwards so didn't find the need of baffles to reduce visual glare. A piece of black felt sits under the ringlight for a black background because little if any LED light reaches it. A magnifying glass stand with lens removed was adopted as the support and suited the 60 mm diameter Petri dish. Its compact form and adjustability for height and angles made it ideal.

I'm delighted with this additional darkfield feature for my stereo studies at no extra cost. Live pond life in particular is striking, both for studying larger fauna and flora and for highlighting tinier organisms to isolate for study under a compound microscope. Samples, especially in water, are more vulnerable supported above the stereo base but with care works well. A stereo with a generous focussing pillar is also needed to focus at the higher focal plane. A seat with adjustable height or a deep cushion ready to hand is also needed to maintain a comfortable posture at the higher eyepiece level.

Compact darkfield condensers for stereos using existing base lighting are readily available—Asian based eBay sellers offer one design at typically ca. £60+ incl.

postage. I've never tried one as they typically cost more than my ringlight and may require high quality stereo base lighting to work well. They also don't have the ready control of the lighting direction and intensity which the ringlight offers. (*With Micscape Editor's hat on, we would welcome readers sharing their experiences on how well these other designs work, or other methods readers use for darkfield stereo work. One or more flexible light guides directed from below and upwards at an angle to the subject is one approach.*)

The YK-B144T ringlight also works very well as a low power darkfield 'condenser' on my Zeiss Photomicroscope III without a condenser or auxiliary lens present—creating velvety black full field backgrounds for objectives from 2.5X up to 10X NA0.22. At the moment a Zeiss achromatic-aplanatic condenser with a darkfield 'D' setting with top condenser lens removed works very well in this role (see article in the October 2008 issue of *Micscape*). But the ability of the LED ringlight to control quadrants may offer possibilities to 'model' subjects under darkfield and something to explore in the future.

Comments to the author David Walker are welcomed.
Acknowledgement. Thank you to Ricardo Y Tsukamoto for highlighting the paper by K F Webb in the Yahoo Microscope Forum, May 22nd 2016.

Above. Freshwater shrimp, (Crangonyx pseudogracilis?) optical mag 25X. The shrimp was very active so ISO 1600 EV-2.0 was used (Sony NEX-5N body, IR handheld shutter remote). No post capture adjustment was made to the black level in the image.

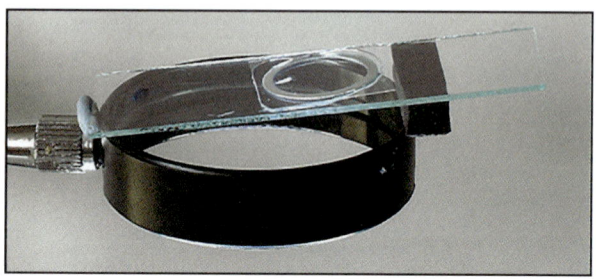

Above. This admittedly rather precarious temporary setup can be useful to maximise the plane of focus for isolated larger specimens photographed under a stereo (e.g. the shrimp above). The slide is raised one end so angled at ca. 6º; i.e. perpendicular to the optical axis of the tube used for photography.

Join a UK Club

Book review: 'Antoni van Leeuwenhoek: Master of the Minuscule'.
by Lesley Robertson, Jantien Backer, Claud Biemans, Joop van Doorn,
Klaas Krab, Willem Reijnders, Henk Smit and Peter Willemsen.
Publisher Brill, The Netherlands, May 2016.

Review by David Walker (UK) from the perspective of a microscopy
enthusiast.
Published June 2016

Image above: Book cover overlain with two commercial high quality replica
microscopes with bi-convex lenses owned by the reviewer.
Left - from the Museum Boerhaave, right - by Chris Kirby, UK of Christopher
Allen Replicas. (The latter with a DIY mount for studying
aqueous samples on a coverslip fragment.)

Additional book details.
ISBN: 9789004304284 (Hardback) ISBN: 9789004304307 (E-book).
Pages XVIII + 229. Extensive colour and monochrome illustrations incorporated in the text.
Pricing: Available direct from the publisher[L1] (link includes publisher's flyer). Hardback, Euros 99 / $128, Ebook $128
(plus tax if applicable), individual E-book chapters $30 each.
Typical UK street price in May 2016. Amazon UK £78.
First published in Dutch as 'Van Leeuwenhoek. Groots in het kleine[L2]' by Veen Media, The Netherlands, January 2014

These are exciting times for those interested in Van Leeuwenhoek's life and work. In the last few years, free online access to two key primary sources have been provided. Three additional microscopes attributed to Van Leeuwenhoek have been reported which has prompted a wealth of fascinating papers and articles. Last February, *Micscape* was delighted to host a careful reassessment by our regular contributor Wim van Egmond of the identity of Van Leeuwenhoek's first reported aquatic microorganism. (See Footnote 1).

Modern biographies of Van Leeuwenhoek in English suitable for a wide audience seem relatively few compared with those for other 17th century scientists such as Newton and Hooke. Clifford Dobell's is regarded as the classic biography (1932). (See Footnote 2 for a selection of biographies in English published to date.)

Antoni van Leeuwenhoek: Master of the Minuscule presents an engaging and copiously illustrated biography in a format which should suit a wide audience and age range. It's the English edition of the biography first published in Dutch by Veen Media in 2014 entitled *Van Leeuwenhoek, groots in het kleine*[L3]. I haven't inspected the Dutch edition but the more recent translation has enabled the discovery in December 2014 of the microscope dredged from Delft canal mud and sold on eBay (misdescribed) to be incorporated. This example is currently being investigated by Brian J Ford[L4] in collaboration with its owner.

The publisher's flyer[L5] notes that "In *Antoni van Leeuwenhoek, Master of the Minuscule*, the Father of Microbiology is presented in the context of his time, relationships and the Dutch Golden Age." and later "This lavishly illustrated biography sets his legacy of scientific achievements against the ideas and reactions of his fellow scientists and other contemporaries."

The scientific and writing backgrounds for the eight authors are provided and all have contributed to a Dutch work on microbiology. Dr. Lesley Robertson's work includes studies with Van Leeuwenhoek microscope replicas and their value in microbiology education. It's not clear from the book how the writing was assigned (possibly by chapter?) but I found the style to be even throughout.

There are twelve chapters as listed below with the pages allocated for each. Within each chapter there are titled sub-sections and interspersed in the text are supporting topics inset in grey boxes. Supporting topics include 'The Connection Between Microorganisms and Illness' and 'The Jealous Rival' [Swammerdam].

1. The Early Years (15pp)
2. Return to Delft (9pp)
3. Antoni's First Brush with Science (21pp)
4. Van Leeuwenhoek's Microscopes (25pp)
5. Antoni van Leeuwenhoek and His Microorganisms (22pp)
6. The Discovery of the "Semine genitali Animalculis" or Spermatozoa (14pp)
7. Antoni van Leeuwenhoek and the Question of Generation (29pp)
8. The Circulation of Blood (13pp)
9. Secrets of Nature (14pp)
10. The Famous Van Leeuwenhoek (20pp)
11. The End of a Long Life (13pp)
12. The Scientific Legacy of Antoni van Leeuwenhoek (19pp)

Chapter 4 includes a valuable illustrated summary of the twelve microscopes currently believed attributable to Van Leeuwenhoek with their magnifications, focal lengths and brief provenance—currently up to date—until perhaps another microscope surfaces on eBay!

Dobell's biography covered his life in detail but discussions of Van Leeuwenhoek's work concentrated on studies of his 'little animals'. These microorganisms are covered in chapter 5 of the present book. Later chapters (6-9) clearly present his important work on spermatozoa, generation and blood circulation in the wider context of earlier and later studies right up to the present day. Chapter 9 provides a flavour of some of the many other subjects he studied including the compound eyes of insects and associated micrometric studies.

In chapters 10-12, a fascinating survey is presented on the impact Van Leeuwenhoek's work had during his lifetime, after his death and the legacy he has left. Legacy aspects include his portrayal on stamps (see also the gallery in the April 2016 issue of *Micscape*[L6]), genera / species named after him and street names.

The authors have found a good balance for the level of depth presented without becoming a dry read. Perhaps the relative merits of the single lens and compound pre-achromatic microscope could have been expanded upon—it is noted that Van Leeuwenhoek's single lenses were superior to the compound microscope optics of the time but not the underlying reasons.

The style chosen doesn't use either in-text numbered references or footnotes and presents a more inviting read than the sometimes dense appearance of a heavily annotated monograph. The 'Selected Bibliography' at the end is kept brief. A potential disadvantage of this approach is that where some authored work is cited, the reader does not have the reference if they wish to read further. Although the extensive bibliography on Douglas Anderson's lensonleeuwenhoek.net[L7] website would allow readers to quickly find them (and cited by the authors as the "best website dedicated to Van Leeuwenhoek at the time of writing.").

The reproduction of the illustrations is to a high standard, presented on bright white paper and incorporated into the body of the text rather than as separate plates. Together with an attractive hardback cover it is a book that is inviting to read. The preview on Google Books[L8] allows a good selection of the first 50 pages to be inspected.

Many of the photomicrographs used in the book have been taken by Dr Robertson using replica microscopes made by Hans Loncke with ground biconvex lenses, including one with a magnification of 303X. Hans has shared his techniques on *Micscape*[L9]. The typical setup she uses is illustrated on her Weblog[L10] on the TU Delft School of Microbiology website. A number of the images are stills from her award winning video shared on YouTube 'Through Van Leeuwenhoek's Eyes[L11]'. I know from my own studies with replicas with only 80-100X lenses that such photomicrography of living subjects is challenging. The authors successfully present imagery (including darkfield) which provides an insight into the sort of views Van Leeuwenhoek may have seen, rather than use modern microscopes and techniques.

The book is commendably free of typographical mistakes and only spotted one. There were a couple of potentially misleading factual points. (See Footnote 3).

The list price of the English edition published by Brill is 99 Euros compared with the original Dutch hardback edition published by Veen Media at 34-50 Euros. I'm not familiar with the factors that govern the price chosen by a publisher, but my feeling is that the English edition is expensive for the readership it is primarily intended for and no saving is offered for the E-book version.

I thoroughly enjoyed reading the book and deserves a wide audience for those wishing to learn more of Van Leeuwenhoek's life and work.

* * * *

Comments to the reviewer David Walker[L12] are welcomed. *The reviewer is a microscopy enthusiast with a particular interest in studying selected subjects which Van Leeuwenhoek reported, e.g. the adult silk moth Bombyx mori[L13], using both modern microscopes and commercial Van Leeuwenhoek replicas.*

Footnotes

1) Recent developments in Van Leeuwenhoek studies:

2011. The Royal Society[L14] provides free online access to its *Philosophical Transactions* dating back to 1665. Many of Van Leeuwenhoek's letters were sent to the Society and translated and published in whole or in part in this journal.

ca.2014. 'The Collected Letters of Antoni van Leeuwenhoek' ('Alle de Brieven ..') - volumes 1-15 of the definitive primary resource now transcribed and available free online on the dbnl.org[L15] website.

2015/6. Papers on the recent microscope discoveries include: *And then there were 12 - distinguishing Van Leeuwenhoek microscopes from old or new copies.*[L16] by Lesley Robertson (2015) and *Genuine or copy? Novel methods of authenticating new Leeuwenhoek microscopes*[L17] by Brian J Ford (2016). *(Thank you to Dr. Lesley Robertson for a copy of her paper.)*

2016 February issue of Micscape.[L18] *The riddle of the 'green streaks'. In search of the first microorganism which Antoni van Leeuwenhoek described. By Wim van Egmond in collaboration with Frans Kouwets. Presents arguments that a coiled cyanobacterium in the genus Dolichospermum not Spirogyra was the first aquatic microorganism described by Van Leeuwenhoek in 1674. (**See Page 38 in this Yearbook**).*

2016 April, published online in Annals of Science[L19], *Antony van Leeuwenhoek's microscopes and other scientific instruments: new information from the Delft archives.* by Huib J. Zuidervaarta and Douglas Anderson. Extensive 32 page paper which includes "... new insights about the way Leeuwenhoek began his lens grinding and how eventually he made his best lenses." (Quote from Abstract.) *(Thank you to Dr. Douglas Anderson for a copy of this paper.)*

2) Selected biographies in English of Van Leeuwenhoek or which cover in-depth aspects of his life and/or work.
Clifford Dobell, 'Antony van Leeuwenhoek and his "Little Animals"', pub. Harcourt, Brace and Company, 1932. Dover reprint, pub. 1960. Available on www.archive.org.[L20]

A Schierbeek, 'Measuring the Invisible World. The life and works of Antoni van Leeuwenhoek', pub. Abel-Schuman, 1959. This is a condensed English edition of his earlier two volume work in Dutch.

Brian J Ford, 'The Leeuwenhoek Legacy', pub. Biopress, 1991.

Edward G Ruestow, 'The Microscope in the Dutch Republic', pub. Cambridge University Press, 1996.

Laura J Snyder, ' Eye of the Beholder. Johannes Vermeer, Antoni van Leeuwenhoek, and the Reinvention of Seeing[L21]', pub. Head of Zeus, 2015. Douglas Anderson has recently had published online in May 2016 an open-access four page review essay on this book in the journal *Studium* entitled 'The tensions between facts and fantasy'.[L22]

3) a) On p.73 when discussing Van Leeuwenhoek's sampling of the Berkelse Meer it states "His examination of the sample while out walking ..." implying that he sampled from the shore. This is translated from the original Dutch letter as "passing lately over this sea" (*Collected Letters*, Vol. I, 1939) and context implying sampling from a boat. (A Dutch colleague notes that the Dutch edition does use "passing" so may have been introduced in the present English translation.)

b) On p.83 the text drawing attention to Figure 5-19 implies that it is an electron micrograph, rather than it being taken with an optical microscope.

c) On p.197 it states that "Single-lens microscopes went out of use in the 18th century, when compound microscopes with at least two lenses - an eyepiece and an objective - became the norm." The single lens microscope continued to have an important role into the 19th century for critical work until the development of achromatic objectives (ca. 1820s-30s) improved the performance of compound microscopes. The single lens designs evolved to models that were more practical with stand, rack and pinion focus and stage. Robert Brown used single lens microscopes for his work on plants and his studies reported in 1828, later to be named after him as Brownian motion. Charles Darwin took a single lens microscope on his *HMS Beagle* voyage.

Published in the June 2016 edition of Micscape.
www.microscopy-uk.org.uk/mag/artjun16/dw-AvLreview.html
09/06/2016

Internet Link References

L1   www.brill.com/products/book/antoni-van-leeuwenhoek

L2   www.boek.be/boek/van-leeuwenhoek

L3   www.boek.be/boek/van-leeuwenhoek

L4   www.researchgate.net/publication/277309807_The_mystery_of_the_microscope_in_mud

L5   www.brill.com/products/book/antoni-van-leeuwenhoek

L6   www.microscopy-uk.org.uk/mag/artapr16/dw-Hooke-AvL-stamps.html

L7   lensonleeuwenhoek.net/

L8   google.co.uk/books?id=qiYiDAAAQBAJ

L9   www.microscopy-uk.org.uk/mag/artjul07/hl-loncke2.html

L10   delftschoolmicrobiology.weblog.tudelft.nl/2016/04/27/playing-with-facsimile-van-leeuwenhoek-microscopes/

L11   www.youtube.com/watch?v=OniSF8QrHac

L12   **email:** micscape@ntlworld.com

L13   www.microscopy-uk.org.uk/mag/artjun12/dw-silk1.html

L14   royalsociety.org/news/2011/Royal-Society-journal-archive-made-permanently-free-to-access/

L15   www.dbnl.org/auteurs/auteur.php?id=leeu027

L16   www.researchgate.net/publication/280057955

L17   www.brianjford.com/a-avl-micanal-1601.pdf

L18   www.microscopy-uk.org.uk/mag/artfeb16/wimleeuwenhoek2.html

L19   www.tandfonline.com/doi/abs/10.1080/00033790.2015.1122837

L20   archive.org/stream/antonyvanleeuwen00dobe#page/n5/mode/2up

L21   headofzeus.com/books/eye-beholder

L22   www.gewina-studium.nl/articles/10.18352/studium.10122/

THE LITHOGRAPHY BUSINESS AT 54 HATTON GARDEN:
FROM OBSCURITY TO CELEBRITY AND PROLIFIC SUCCESS

By Peter B. Paisley
Sydney, Australia
Published May 2016

1844-52: A NEW HATTON GARDEN ENTERPRISE

Prior to 1844, 54 Hatton Garden, Middlesex, probably contained printing apparatus (including lithography equipment) belonging to Robert Hone. On 30th January 1844, the *London Gazette* announced bankruptcy proceedings against him (below).

EDWARD HOLROYD, Esq. one of Her Majesty's Commissioners authorised to act under a Fiat in Bankruptcy, bearing date the 30th of January 1844, awarded and issued forth against Robert Hone, late of 27, Garnault-place, Spafields, but now of No. 54, Hatton-garden, both places in the county of Middlesex, Stationer, Dealer and Chapman, will sit on the 11th of June next, at twelve at noon precisely, at the Court of Bankruptcy, in Basinghall-street, in the city of London, to make a Dividend of the estate and effects of the said bankrupt; when and where the creditors, who have not already proved their debts, are to come prepared to prove the same, or they will be excluded the benefit of the said Dividend. And all claims not then proved will be disallowed.

By 6th February, bankruptcy was formally confirmed, witness a notice in the *Legal Observer* (vol.27, p.367), below.

Hone, Robert, late of 27, Garnault Place, Spa Fields, but now of 54, Hatton Garden, Stationer. *Edwards.* Off. Ass.; *Catlin*, 39, Ely Place, Holborn. Feb. 6.

Presumably 54 Hatton Garden and its contents were disposed of by the court in June 1844, as scheduled. Despite long searching, I found nothing more about Robert Hone. His double description as "dealer and chapman" (if not tautological) implies both manufacture and sale of pamphlets and small books – hence printing equipment. Whatever the case, by 1845 or shortly thereafter, new occupants were at 54 - the lithographers Ford and George – a partnership which lasted until 1852. George Henry Ford yields an abundance of evidence both on his own and in collaboration with others: B.G. (Ben/Benjamin George) George, whose forenames and initials are often misquoted, yields much less; and Hone – judging by internet sources - has sunk into total obscurity. (Readers may know something about him – I would welcome evidence.)

Ford and George made lithographs featuring a wide variety of subjects, canvassing business in several books and periodicals. The prime mover was George Henry Ford, who had a firmly established reputation for first rate natural history illustration, well before his move to Hatton Garden. One gets the impression he was "treading water" with general popular lithography while seeking medium or long term contracts in biomedical and/or natural history publication. The advertisement (below) in *Medical Times* of Saturday 19th May 1849

indicates his preference by the chosen periodical in which it appeared, and by emphasising "scientific".

Lithography. — G. H. Ford begs to acquaint his Friends, and Gentlemen engaged upon Scientific Publications, that he now carries on business as a GENERAL LITHOGRAPHER, in Partnership with M. B. George, at No. 54, Hatton-garden, where he will be happy to be favoured with orders for any variety of Lithographic Drawings and Printing, either Scientific, Architectural, or Picturesque, which will be executed in the first style of art, and at very moderate prices.

At the same time, the need for income stimulated continued production of non-scientific lithography, evidenced by the *Medical Times*' mention of varied subjects (and the offer of competitive prices). In addition to "scientific, architectural, or picturesque" subjects, portrait lithography of military heroes, widely read poets, and heads of learned societies often found favour with the general public.

The "picturesque" could be enhanced by far-flung geographical appeal – from New Zealand for instance, and Ford and George provided some lithography for *An Account of the Settlement of New Plymouth* (1849) by Charles Flinders Hursthouse. Colonial scenes were popular, and agreement may have existed with the publishers to issue single sheet lithographs after the book's appearance. Examples of the New Plymouth illustrations are shown below.

"Glenavon Farm", Capt. Davy's residence.

"Brooklands", residence of Henry King RN

Hursthouse's preface is dated September 1848. Books take time – letterpress is typed, then proofs checked, illustrations reviewed, and so on: the lithography was probably done in 1847 or even 1846, so it is reasonable to assume that Ford and George had settled into 54 Hatton Garden by then, or perhaps as early as mid 1845, given time taken to import and arrange their equipment.

Portraits could be based on photographs like platinographs, or daguerreotypes like that of Edward Doubleday, below.

Doubleday died in 1849, and had worked at the British Museum – as had Ford, who therefore doubtless knew him.

Advertisement in the "Athenaeum",
December 29th 1849

A PORTRAIT of the late EDWARD DOUBLEDAY, Esq. F.L.S. F.Z.S. of the British Museum, is being Lithographed by G. H. FORD, from a Daguerreotype by J. W. GUCH, Esq. and will shortly be ready for delivery. Price, Artists' Touched Proofs, 10s.; India Proofs, 8d. Only 180 Impressions will be taken, 50 of which are already subscribed for. Subscribers' names received by Messrs. FORD & GEORGE, Lithographers, 54, Hatton-garden; and by Mr. HOGARTH, Printseller, Haymarket.

Somewhat unusually, the *Athenaeum* advertisement above gives precise differential price details. By making the Doubleday portrait a limited edition, Ford and George used sales psychology to enhance business. The Doubleday print could be bought directly from Hatton Garden, or via the well known retail shop of Hogarth. Without access to sales records, one can only guess what percentage went to retail shops: no doubt, the higher the repute of the lithographer, the less reliance there was on secondary outlets. Once again, for those buying direct from Hatton Garden, the firm has a sales pitch – "prices more moderate than are usually charged".

FORD & GEORGE, LITHOGRAPHERS, 54, HATTON GARDEN, beg to announce that they execute in the best manner every variety of Lithographic Drawing and Printing, both plain and in colours, including Portraits, Views, Architectural and Picturesque; Anatomical, Geological, and Animal Subjects; Plans, Maps, Music-Titles, and every description of Plain and Ornamental Writing, at prices more moderate than are usually charged.—Specimens and Estimates of Work in a superior style, or of a more economical description, furnished when required.

Athenaeum advertisement, May 1849

There was a commercial balance between selling via retail galleries and selling direct: retailers charged a premium, but might enhance lithographers' reputations by exposure to a wide clientele. The Colnaghi shop, especially, attracted members of the aristocracy – even royalty. Ford was well known in natural history circles, but Ford and George prints of a widely varied nature needed exposure to a public less likely to visit Hatton Garden than to attend fashionable gatherings at Colnaghi's gallery. There is an obvious parallel with sales of microscopical mounts. Consider the multitude of surviving mounts by John Barnett, for example. Judging by my collection and offerings at auctions, his mounts lacking optical shop labels greatly outnumber those bearing them. Having established a reputation, Barnett sold from home, using word of mouth recommendations, or free publicity in magazines like *Science Gossip*.

As the notice below shows, a considerable price was charged for large prints: depending on how many were sold, the portrait of Field-Marshal Sir George Pollock may have brought in a tidy sum for the time.

POLLOCK (Maj.-Gen. Sir Geo.).—In Lithograph, by G. H. Ford; size, 15 inches by 20 high; Prints, 10s 6d. Colnaghi.

Two Barnett mounts from my collection

As the notice below shows, a considerable price was charged for large prints: depending on how many were sold, the portrait of Field-Marshal Sir George Pollock may have brought in a tidy sum for the time.

POLLOCK (Maj.-Gen. Sir Geo.).—In Lithograph, by G. H. Ford; size, 15 inches by 20 high; Prints, 10s 6d. Colnaghi.

By selling via Colnaghi, Ford and George came to the attention of the wider press, witness the Spectator notice, (see over).

Notice in the Spectator, 7th September 1850. Below: more portraits: (L) the poet Leigh Hunt, (R) General Pollock.

Guide books, street maps and architectural vignettes were other vehicles for Ford and George lithography, all combined in a London miscellany published in 1851.

The book contained many street maps, which could perhaps be purchased singly from Hatton Garden. It also contained the firm's most flamboyant advertisement, occupying almost a full page (right, Fig. 1).

Through the 1840s, Ford and George doubtless continued to sell lithographs – picturesque views, portraits, maps and so on: but on 30th December 1851, the London Gazette foreshadowed winding up of the partnership's affairs. (See Top Right, Fig 2.)

Fig.2

NOTICE is hereby given, that the Partnership subsisting between us the undersigned, George Henry Ford and Benjamin George George, as Lithographers, Printers, and Colourers, under the style or firm of Ford and George, carrying on business at No. 54, Hatton Garden, in the county of Middlesex, has been this day dissolved by mutual consent. All debts owing to or by the said partnership will be received and paid by the said George Henry Ford.—Dated this 30th day of December 1851.

G. H. Ford.
Ben. G. George.

From the above notice one can conclude that the partnership ended amicably. George preferred to continue lithography of widely varied subjects, as opposed to the technical scientific direction which Ford preferred. On January 1852, The Economist reprinted the London Gazette confirmation of their partnership as formally dissolved (below).

singhall Street, attornies—Trendell and Bracher, Reading, watch-makers—Ford and West, Hatton Garden, lithographers—Hawksley and Sturt, Margaret Street, Caven-

George moved to his own premises at 47 Hatton Garden, and if "snippets" from Google Books are any indication, he flourished there, with employees, as indicated in the C.V. fragment below:

Congressional Edition - Volume 5556 - Page 8294
https://books.google.com/books?id=tvVGAQAAIAAJ
1909 - Snippet view - More editions
From November 8, 1907, to January 14, 1909, I have been employed by the Providence Lithograph Company, of the city of ... deposes and says: I am a lithograph artist and was employed as such by the firm of Ben George & Co., at London, ...

THE BRITISH METROPOLIS ADVERTISER.

Fig.1

FORD AND GEORGE,

LITHOGRAPHERS AND ENGRAVERS,

54, HATTON GARDEN, LONDON.

Architectural and Pictorial Drawings executed by Artists of the first talent; Original Designs for Music and Book Titles, and Ornamental Covers of an elegant and attractive kind, in Gold and Colours, will be furnished when desired.

Maps, Plans, Circulars, and every description of Plain and Ornamental Writing for Engineering and Commercial purposes executed in the most finished style.

F. and G., having all requisite facilities for the speedy completion of every description of Lithography, can confidently insure against disappointment in respect to the punctual completion of orders in all cases, more particularly those requiring extraordinary dispatch; and their Scale of Charges will be found to bear advantageous comparison with that of any other house.

"Ben George & Co." and "Ben George Ltd." indicate some prosperity, which creates goodwill for any business: another Ben George acquired the firm's goodwill in1904. I assume this other Ben George was a family member, perhaps a son (below).

The firm had sufficient funds to apply for a US patent in 1876, possibly for glazing of tin biscuit boxes done in London since 1868.

This probably refers to decorated tin sheets described by the Victoria and Albert Museum (below), providing material for Huntley and Palmers, which by around 1870 had become a giant multinational company with a global export reach extending to the USA.

Examples of Benjamin George's later work:

Above: a Huntley and Palmers biscuit tin, with transfer printing by Owen Jones and finishing by Ben George &Co.

Right: caricature of Henry Irving as Richard III

So much for George after his split from Ford and George: his different interests are clear. (En passant, Google Books' treatment of us colonials as third class citizens is plain from the above: letting us read

more than snippets perhaps threatens world peace.) Obviously, Ford's preference was for the technical, while George's was for the decorative: besides, by now Ford had probably met Tuffen West, who was well on the way to becoming another celebrity in biomedical illustration. Regardless of this, Ford had probably long since developed a taste for the microscopical, given the eminent biomedical authors with whom he had worked. By the middle 1850s, judging by his work for James Samuelson (of which more below) he was already familiar with the microscope and what it could reveal.

Before the partnership ended, Ford and George had begun lithography related to Ford's scientific interests: by chance, the New Plymouth views, seen on a National Library (NZ) web site, led to something which (I trust) is more interesting to Micscape readers.

Geology

As well as the two New Plymouth views, the NZ library catalogue mentioned another Ford and George lithograph, from volume 6 of the Quarterly Journal of the Geological Society (London), which led me to discover more such plates in the same journal in that year.

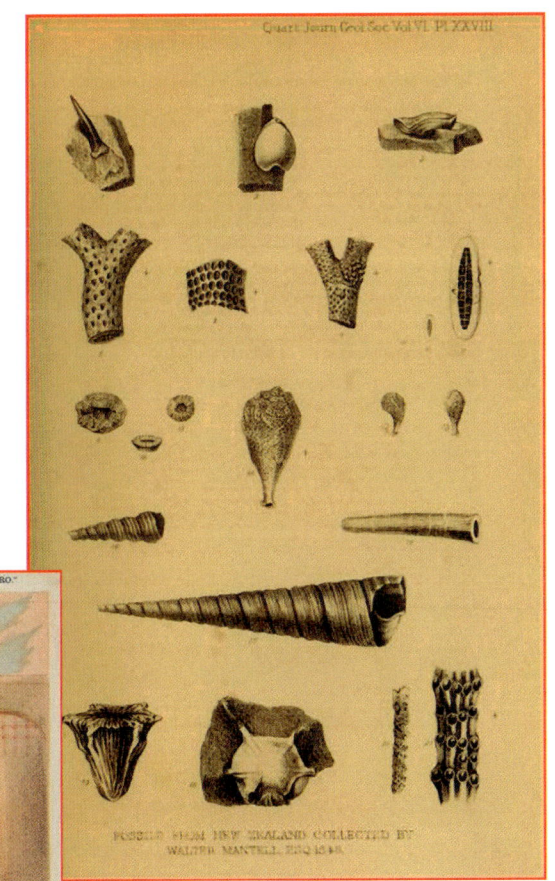

The plates above (and next page, top left) illustrate a paper based on fossils collected in New Zealand by Walter Mantell, read on 27th February 1850 by his father Gideon Mantell.

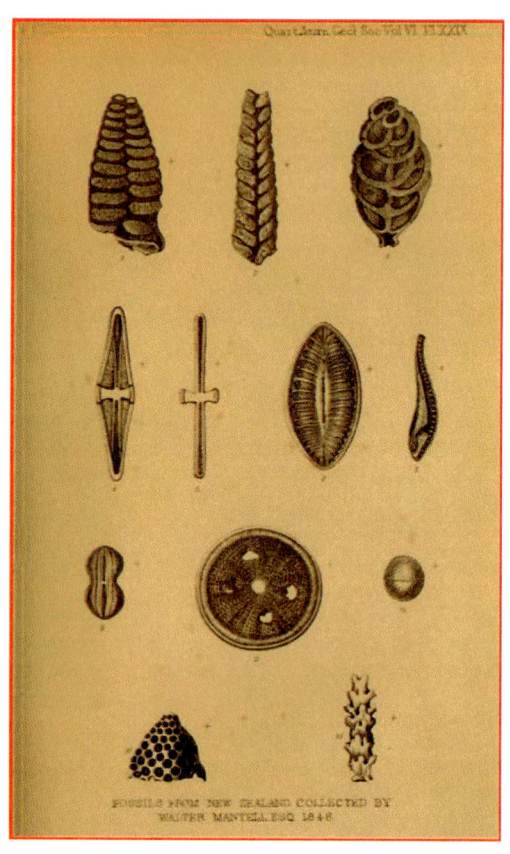

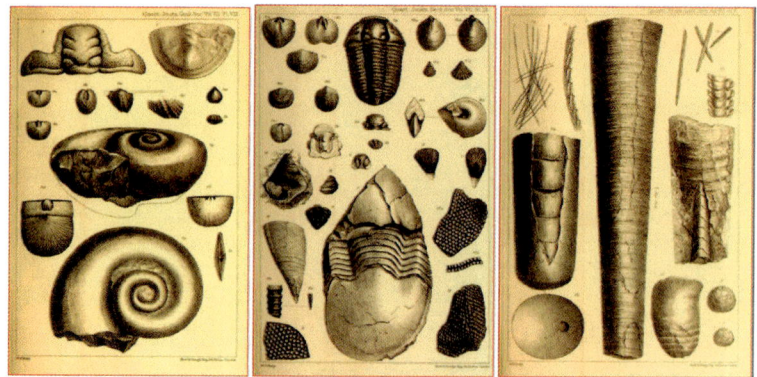

Three plates illustrating a Murchison paper on Silurian rock fossils, 25th February 1851

Walter Mantell almost certainly discussed the Hursthouse illustrations of New Plymouth with his father, probably drawing his attention, and that of Murchison, to the Hatton Garden firm.

Reading further in that volume (via archive.org) I found more examples. Ford and George provided only a few lithographed plates for early Geological Society proceedings, but these "new boys" were noticed by Roderick Murchison, who attended most of the same meetings as Mantell. In the early to mid 19th century – the age of superstars of geology – none were more stellar than Mantell and Murchison, so anything from 54 Hatton Garden was looked on with considerable interest.

Below: on 18th April 1849, Ford and George had provided lithography in the *Quarterly Journal*, illustrating fossil fish for a paper by Egerton.

Later, on 11th June 1851, they once again provided lithography, for a paper by Sykes and Egerton, on a fossil fish from India (below).

Scientific illustration: the Cinderella of art history

Recently I asked a world renowned art historian (with a strong interest in science) how to find scientific engravers, lithographers and printers. The reply:

authors would not mention such details. But I began reading book prefaces, which usually name and sometimes lavishly praise the artists. So much for mainstream art history!

With journals, the task is harder. Editors rarely mention illustrators, so only fine print at the bottom of plates reveals them; but digitised quality is uneven, from superb (often with *Proceedings of the Zoological Society of London*) to appalling (distressingly often with other publications). Ignoring the value of illustrations creates shoddy work - plates blurred, illegible or absent, with no trace of fine print. Mercifully, enough clear examples exist to enable an historical narrative.

Other sources exist, but many universities now demand outrageous prices for scans, and the *historical directories* web site, "improved" by geeks to justify their salaries, is now impenetrable. Such sources once provided corroboration, at times with "stand alone" facts absent from databases like ancestry.com – alas, no more.

Google Books, if useful, is a two-edged sword: we colonials often get mere "snippets". Penetrating the censorship may stimulate others to try: one day, any knowledge generated should join mainstream art history.

Winds of change: multiple artistic collaboration at 54 Hatton Garden

While still in partnership with Ben George, Ford already collaborated with William West (Tuffen's younger brother). Exactly when William arrived in London seems undocumented, but by 1853 he and Ford, having formed a partnership, had created scenic lithographs. Hence, his first occupations were both printing and lithography, either on his own or with Ford at 54 Hatton Garden. Probably through contacts made during the Ford and George partnership, and their New Zealand lithographs, they printed the Canadian views below.

Above: York Factory.
Below: buildings seen from across a river bend

In the context of William West's work as a whole, these exercises in the "picturesque" are atypical: the partnership did not last long, and was gazetted as dissolved on 8th May 1853.

While the formal business association had been wound up, the relationship did not end there, and co-operation – in chromolithography – continued for some years afterwards. Probably under the influence of William and his brother Tuffen, scientific subjects soon supplied material, and examples are shown below. I selected these from plates in the *Proceedings of the Zoological Society of London*: that society took great care with its illustrations, and (for once!) the digitised internet versions are uniformly excellent. Plates were issued in separate volumes as well as in the articles they illustrated, making internet browsing easy.

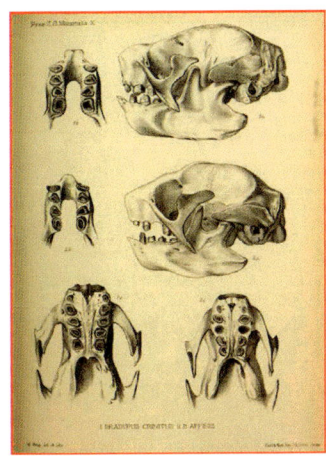
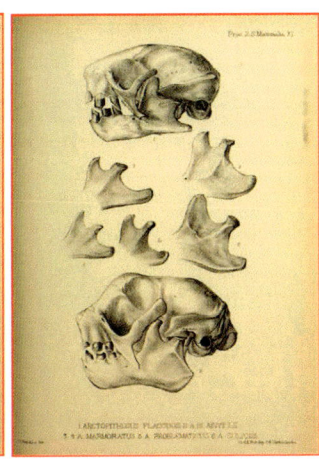

Above: three plates from the mammalian series, marked "Ford and West 54 Hatton Garden", all originally published in 1849. The deer plate is hand coloured.

Below: another Ford and West printing, from 1849, this time from the reptiles series: it does not specify 54 Hatton Garden, and may have been done elsewhere.

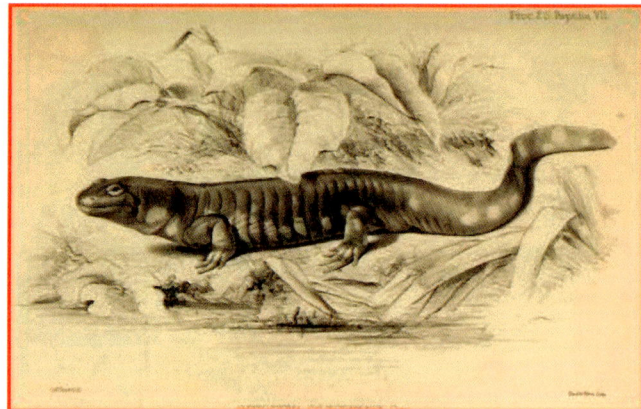

William West was getting opportunities to work with important natural history artists – not just Ford, but others - William Wing and Joseph Wolf, for instance, who were regular colourists for researchers associated with the British Museum - Wing engraved the left skull figures above, and Wolf was one of a team of half a dozen artists who hand coloured John Gould's plates.

Papers take time to prepare, and most journals have a publication delay due to peer reviews and pressure of author numbers, so 1849 may be at least a year after the plates above were prepared, and the original engravings may date from 1848 or even late 1847 – some four years before the Ford/George partnership was dissolved.

Microscopy

I found no evidence of microscopical work by Ford prior to the middle 1840s. Nevertheless, he had been involved in natural history for many years, and must have been aware of rapid advances in microscopy: besides, if he did not already know of Tuffen West's work, he certainly would have by around 1847, through working with Tuffen's brother. Certainly by 1848-9 he must have seen William West's printing of William Wing's lithograph, published (1849) in the *Proceedings of the Zoological Society of London.*

Hippopotamus skin scrapings.

The illustration includes the information (L) "W. Wing del et lith"and (R) "W West imp."

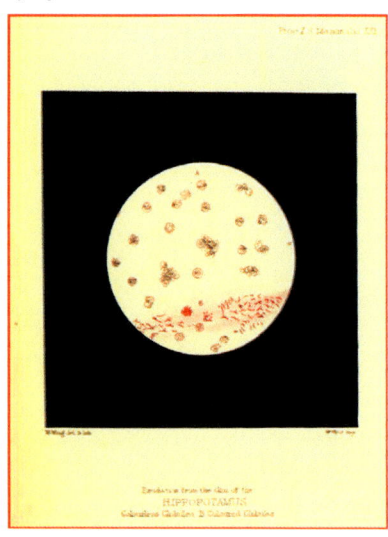

William West, along with his brother Tuffen, probably whetted Ford's appetite for microscopical illustration: whatever the case, later in the same decade Ford was creating lithographs of microscopical material, which William West printed. For instance, the Ford/West partnership attracted the attention of George Viner Ellis, Professor of Anatomy at University College London. They were chosen to make two plates for Ellis' paper on microscopy of bladder involuntary muscle, read at a Royal Society meeting in December 1858.

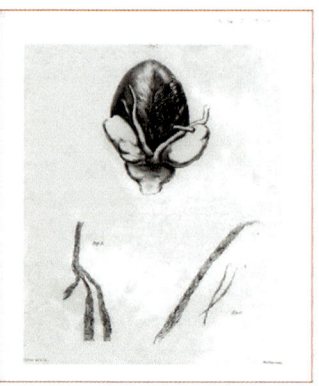 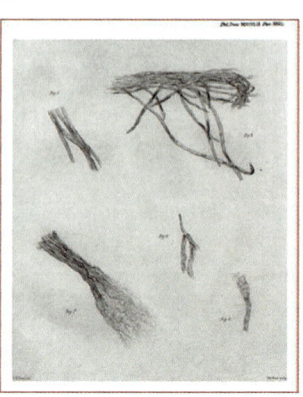

Plates from Ellis' paper for the Philosophical Transactions of the Royal Society.

As I shall show presently, this was by no means the last time Ford and West shared a project for George Viner Ellis, well after the partnership was legally dissolved in 1853 (a major collaboration occurred during the 1860s.).

James Samuelson wrote two short books on "humble creatures", firstly on the earthworm and the house fly (which had two editions, 1858 and 1860) then on the bee (1860): both books described microscopy of these "humble creatures", with G.H. Ford providing plates. Several digitised versions of these books are on the internet: sometimes the standard of the illustrations is atrocious. In the bee book, the plates were tinted (i.e. probably hand coloured by Ford) but in none of the digitised versions has this been reproduced.

Below: frontispieces by Ford (L) for the earthworm/fly book, (R) for the bee book.

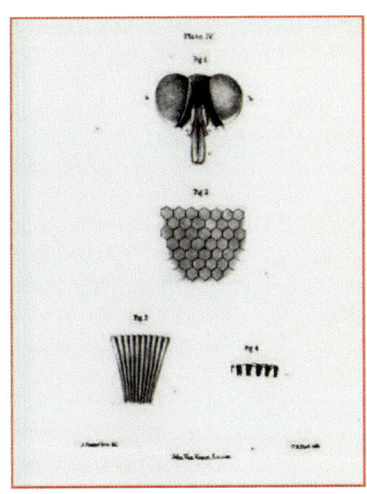
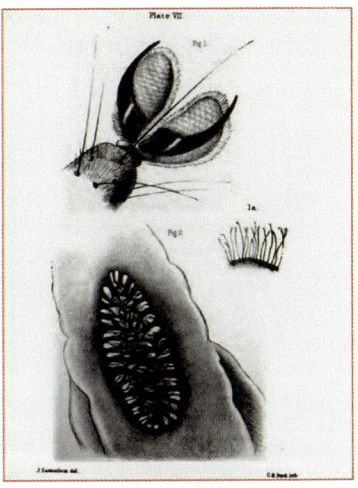

Left: Ford lithographs of fly microscopy

Below: Ford's lithographs of bee microscopy

Bottom: Below: Ford's lithographs of bee microscopy

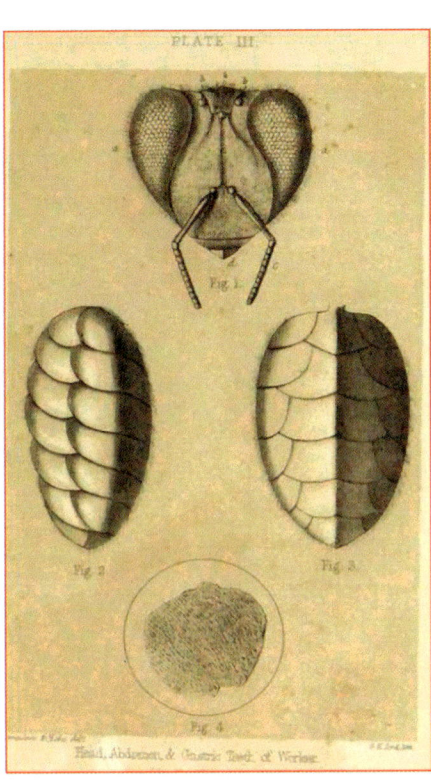
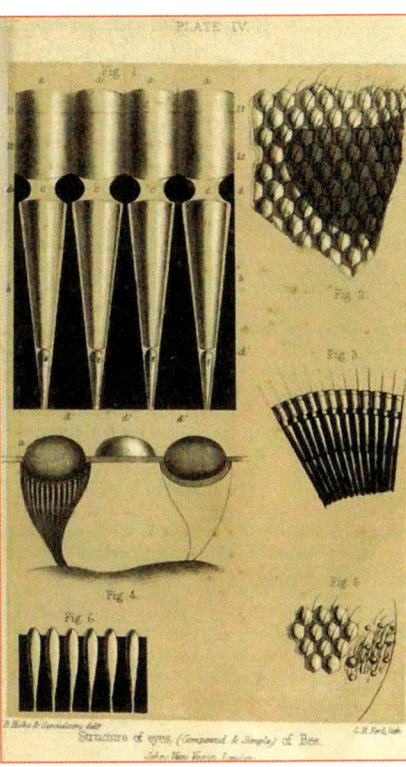
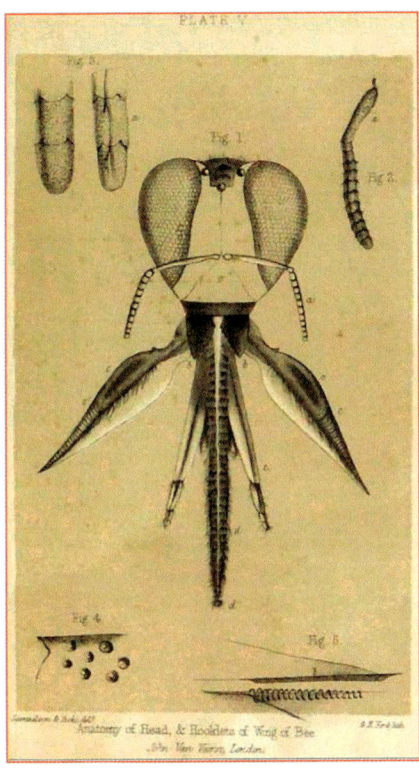

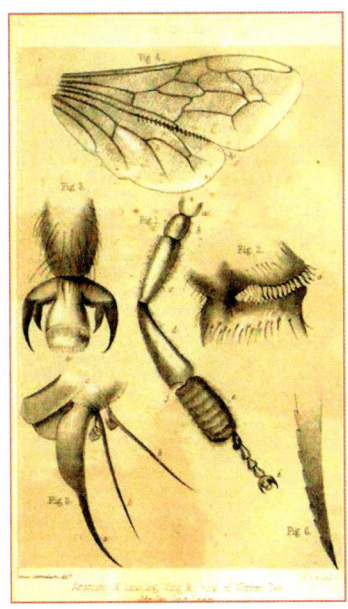
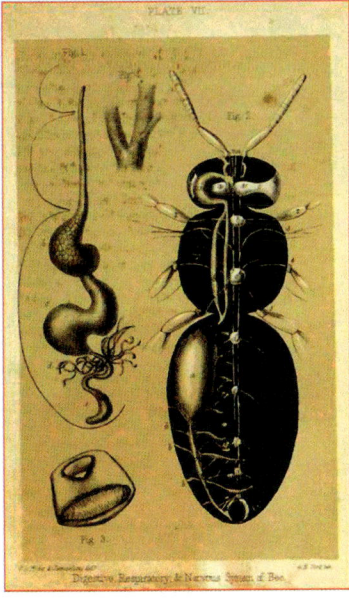
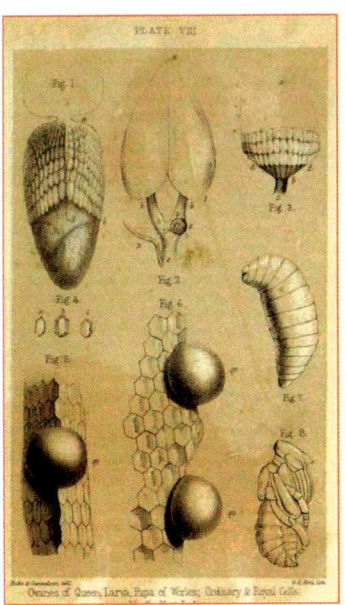

A new quarterly periodical, edited by James Samuelson – *The Popular Science Review* – first appeared in 1862: Ford was among its illustrators, and William West was de facto its resident printer. The most frequent legend at bottom right of the plates is "W. West imp." Microscopical material was not always featured, but usually it was: likewise, William West was not always the printer, but almost always he was. Eight plates printed by him are shown below.

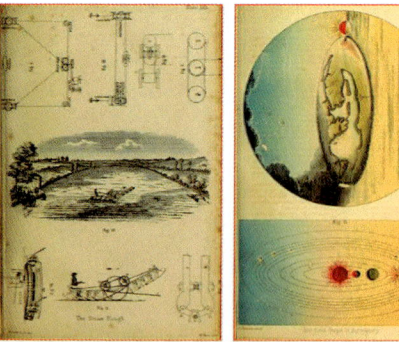

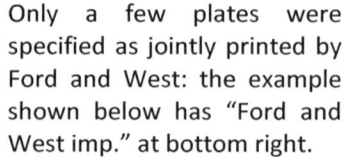

Usually but not invariably William's plates involved microscopy: sometimes he had an opportunity to print other subject matter, as in the two examples left, from articles on the steam plough and primitive astronomy.

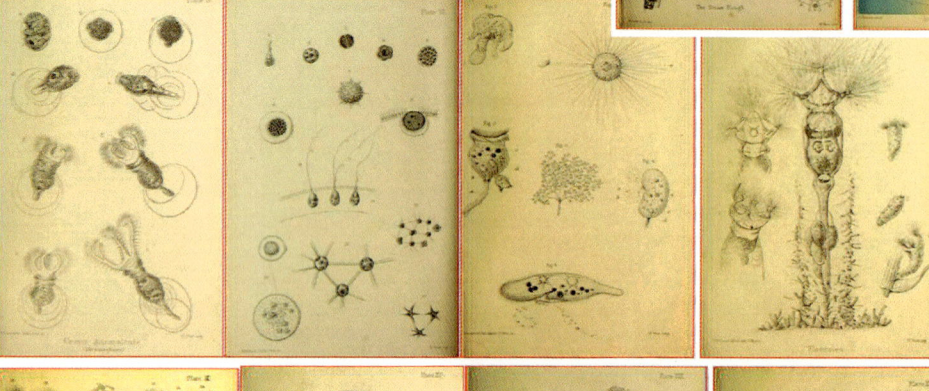

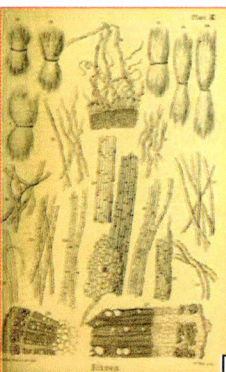
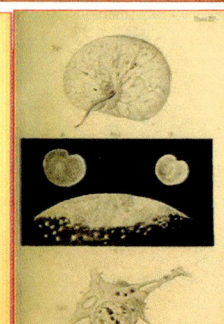

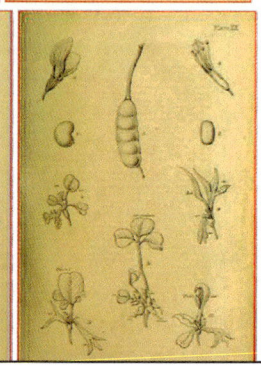

Only a few plates were specified as jointly printed by Ford and West: the example shown below has "Ford and West imp." at bottom right.

Below: three of the many plates printed by "W. West and Co." in "The *Popular Science Review*": (L) in 1874 (middle) in 1876 (R) in 1878. William West died in early 1870.

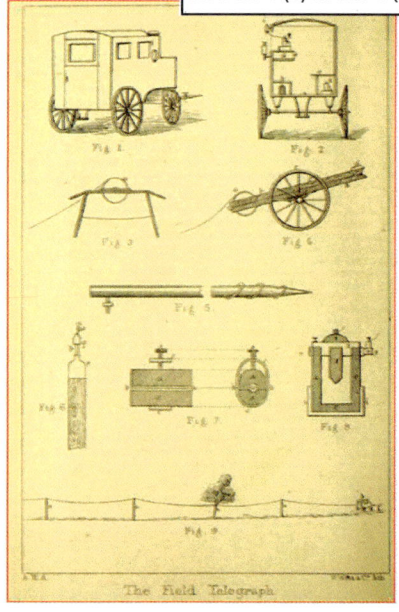
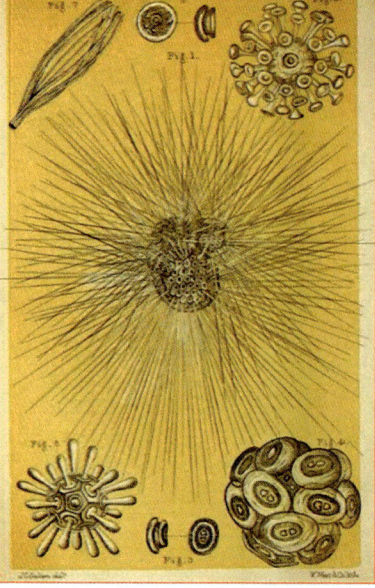
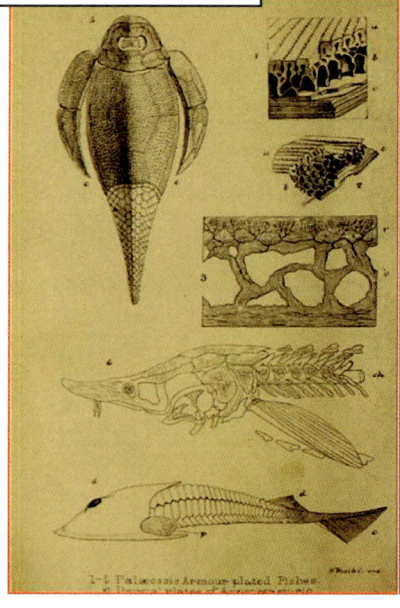

If The *Popular Science Review* is typical, William West shouldered the bulk of the printing activity at Hatton Garden during the 1850s. That activity was nothing short of prodigious, since for a decade after William West's death in 1870, material printed by him continued to appear in *The Popular Science Review*, and elsewhere.

The three plates above raise interesting questions:

chiefly, who was the "and Co." of "West and Co."? The material shown in the middle of the three is from the *Challenger* expedition, therefore could not possibly have been printed by William West. So - what is the explanation? Many examples of "W. West" plates appear in publication for a year or two after William's death, and almost certainly represent stock in hand at Hatton Garden, in a queue for publication. I strongly suspect that the "and Co." is William's brother Tuffen, and possibly other family members – William's wife, for instance (although she pre-deceased him by almost a year). But there are indeed examples of genuinely posthumous William West printings, particularly in partnership with his brother Tuffen: these presumably were still in queues for publication at the time of William's death.

Throughout most of the 1860s, G.H. Ford was making many illustrations. He was responsible for several in Charles Darwin's *The Descent of Man*: an example is shown below.

Both William's and Ford's time throughout the middle 1860s was significantly occupied in their work for George Viner Ellis, who followed Richard Quain as Professor of Anatomy at University College London. Ellis continued an unbroken London series of large anatomical illustrations, started by Jones Quain and continued by his brother Richard, both from University College. These were regional dissection depictions, on page dimensions which allowed for life size illustration. Ford, of course, was long since celebrated for his hand coloured plates for Andrew Smith's *Illustrations of the Zoology of South Africa*. Ford had accompanied his father to South Africa in 1820: on his return to England in 1837 he was recognised as a major talent, and found work at the British Museum.

Below are four of Ford's hand coloured South African illustrations.

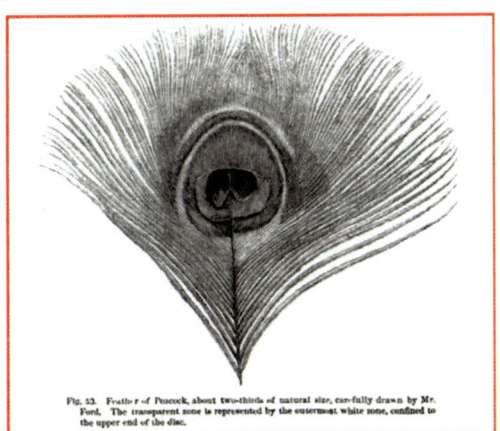

Fig. 53. Feather of Peacock, about two-thirds of natural size, carefully drawn by Mr. Ford. The transparent zone is represented by the outermost white zone, confined to the upper end of the disc.

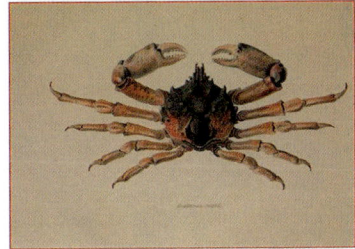

George Viner Ellis' regional anatomical dissections (compiled between 1864 and 1867, when they finally appeared in book form) displays a radical stylistic shift from Ford's other work. The plates, specifically aimed at surgeons in training, are strongly reminiscent of those in Joseph MacLise's surgical anatomy illustrations from some eight years earlier.

Images right (U/L) From the MacLise book (U/R) From the Ellis book. (L/L) from MacLise (L/R) from Ellis.

An interesting similarity between the two sets of illustrations is that both MacLise and Ford sign their work in cursive writing, as opposed to having their names printed at the feet of the plates. William West, quite

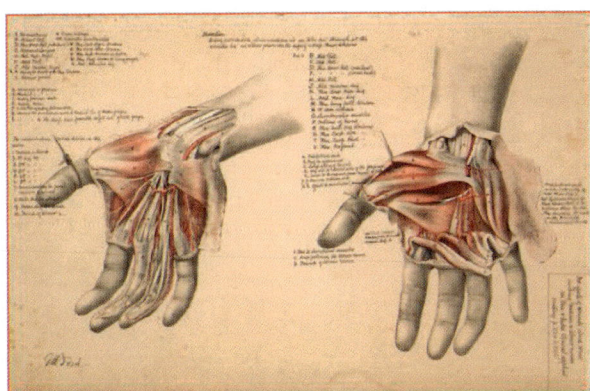
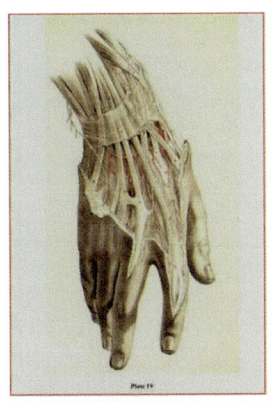
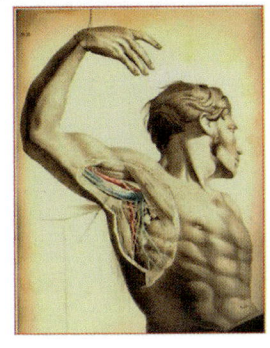
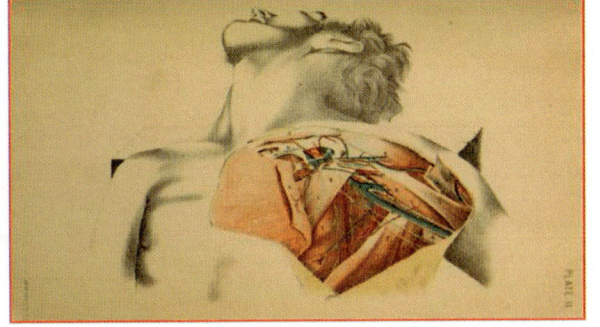

unusually, gets special praise for printing, in the preface to George Viner Ellis' book:

"Before closing this retrospect of the task now finished, I may advert to the difficulties attendant on the printing in colours of such complicated Figures, and to the successful way in which they were overcome by Mr. West."

As always seems the case when Ford and West worked together, chromolithography was the favoured method. Ellis' book, through several editions, was reprinted many times on both sides of the Atlantic: the latest digitised version I saw was from 1891 – and praise for William West as its printer remained unaltered throughout.

Competition with lithography and within lithography

In the mid 19th century, photography – as an art form - competed with painting and lithography. It often claimed "reality", with public appeal as a "true" record of contemporary events. The first newspaper claiming such accuracy was *The Illustrated London News*: but photographs could, and did, lie. Carl Popper says somewhere that there is no such thing as history without a point of view, and the same was now true of the view itself. In *The Illustrated London News*, art was enlisted as propaganda in the Crimea, where rivalry between photography and lithography produced sharp contrast. There were two war correspondents – Roger Fenton, who took photographs, and William Simpson, who made water colour sketches which were converted to lithographs and sold by Colnaghi. Fenton is known to have "doctored" several of his photographs, before conversion to wood engravings for publication, with details further altered in the process. Things were romanticised, as in the picture below, where there is no trace of festering wounds, missing limbs, typhus, or typhoid and Florence Nightingale holds centre stage like a pre-Raphaelite Madonna. Posthumous accolades, indeed!

Equally romanticised is the paper's woodcut of the charge of the Light Brigade, portrayed as orderly British valour - which differs from William Simpson's well known lithograph of the same event, which shows the charge from the Russian side, doing more to portray it as the incompetent military madness it actually was.

(Above) What the Illustrated London News thought its readers wanted to see:.

(Below) William Simpson's lithograph. Both versions, of course, show disciplined military formation, not the shambles which actually occurred.

There were other pressures on hundreds of London lithographers competing for business: new methods emerged, notably the "lithotint" method pioneered by Charles Hullmandel and favoured by many journals, as well as by the influential Colnaghi gallery. Hullmandel died in 1850 but his method continued to flourish, and it offered more subtle shade variations than the more traditional "chromolithography", used by William West and G.H. Ford. (Most of Ford's work was hand coloured by himself: however, when working with William West, they favoured chromolithography.) When

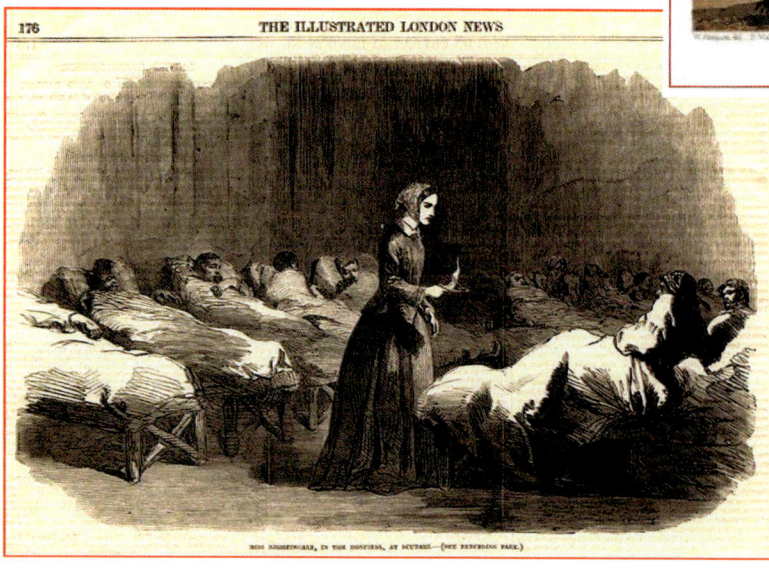

William worked with his brother Tuffen, their output, rare among full time illustrators, was grounded in solid knowledge of much of what they illustrated, namely microscopy. Unsurprisingly therefore they were constantly in demand by biomedical authors. An additional edge for the Wests (and others) was that until the late 19th century, printed microphotography was expensive, so that lithography and/or wood and metal engravings offered cheaper alternatives to publishers. Hence, those, and not photography, dominated biomedical literature, particularly in matters microscopical, long after photography became easier and cheaper – but not as part of letterpress. One way of reducing costs was for an author to make and paste his or her own photographs: this however was cumbersome, and only practicable for fairly small print runs.

This book (see right) – by a past president of the Quekett Microscopical Club – used the author's own photographic prints throughout. (The first edition appeared in 1887.)

Unless publishers were confident of large sales and high prices, integration of photography with text only came into its own late in the 19th century, so lithography and wood engraving long continued to hold sway.

No matter what methods were used when it came to illustration of microscopical material, it might seem that sketches made with the camera lucida would guarantee accuracy, and any colouring would be limited only by the range of pigments available. Nevertheless, illustration was inevitably influenced by theory (or wishful thinking). Between specimen and illustration, there was opportunity for misinterpretation. Distortion from fixation techniques might occur; limitation in lens technology could add a further layer of uncertainty; inaccuracy in drawing or tracing might produce what seemed to be there, as opposed to what actually was; engraving, be it on wood, steel or copper, allowed further opportunity for distortion; composite lithography combining figures from engraving added yet more possible inaccuracy. Finally – and this usually attracted little comment – printing played an important part in what finally appeared for critical appraisal.

Psychological factors were, potentially, all too omnipresent. As Carpenter warned:

"It is a tendency common to all observers, and not by any means peculiar to Microscopists, to describe what they believe and infer, rather than what they witness."

(The Microscope and its revelations, 1856)

That tendency might be reinforced by any of the processes along the way to publication: the final stage – printing – seldom excited criticism. But, as Ellis recognised, it was no less vital than any of the others. William West is the brother most overlooked by historians, but his part in biomedical illustration is not to be underestimated.

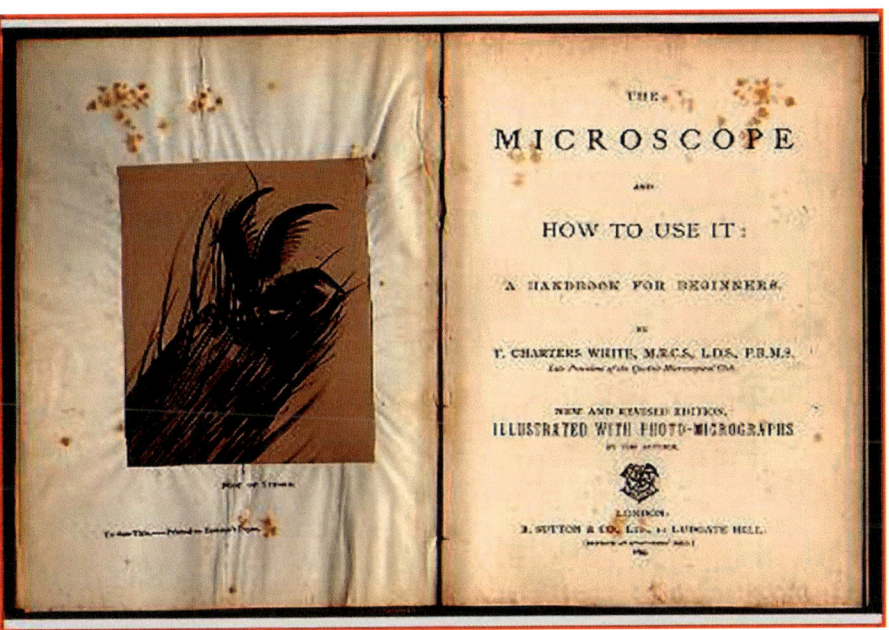

Acknowledgments

Thanks to colleagues who have provided constructive criticism (they know who they are!)
Sources: Google Books, www.archive.org, Wikipedia

Contact author Peter B. Paisley at:
lois737@bigpond.com

Editor's note:
The history of microscopy, especially where it relates to recreational microscopy practised by wealthy Victorian families, is a highly regarded aspect of Enthusiast Microscopy. It is a subject followed by many individuals in the Microscopical community who often privately research previous contributors to the pursuit. Our online presence at **www,microscopy-uk.org.uk** and **micscape.org** carries many such articles, well researched.

Reflections on studying *Spirogyra* - a classic school biology subject and plenty of interest for the hobbyist.

by David Walker, UK
Published February 2016

One of the more distinctive filamentous algae is *Spirogyra* with its spirally arranged chloroplasts. For the microscopy hobbyist it offers plenty of interest and Micscape contributors have shared a selection of articles (see Related Micscape Articles section). This article concentrates on the aspects below from my own recent studies:

1) *Spirogyra* under a commercial Van Leeuwenhoek replica microscope. *My own interest in this algae was piqued when helping Wim van Egmond share his February 2016 Micscape article [The Riddle of the 'green streaks'. Antoni van Leeuwenhoek: In search of the first microorganism he described](N1) (Page 33). In this article he reassesses whether Leeuwenhoek did first describe Spirogyra in his letter to the Royal Society dated Sept. 7th 1674. Wim, in collaboration with a phycologist colleague Frans Kouwets, presents persuasive arguments that the later attribution of Spirogyra was not the most likely candidate—another organism matches the features in Leeuwenhoek's description much more closely. We agreed that it would be a useful complement to share images of Spirogyra taken through a replica Van Leeuwenhoek microscope.*

2) An admittedly self-indulgent sharing of my school level studies.

3) The usefulness of *Spirogyra* for exploring different lighting techniques including autofluorescence with simple filter additions to a typical transmitted

Above, June 2004 photograph. Water troughs hewn out of the local millstone grit (a coarse-grained sandstone) are an interesting upland habitat in the north of England where I live (see my June 2004 Micscape article). Some, like this spring fed trough, could be regarded as a slow moving river with its continuous water flow for most of the year. They also offer varied habitats in a small area, e.g. splash zones with bryophytes, damp mud, open water, rocky substrates etc.

compound microscope with darkfield facilities.

4) A good resource for attempting identification to species in Britain.

5) Some typical commercial prepared slides, including the set offered by the late Eric Marson of Northern Biological Supplies (NBS) showing conjugation.

Spirogyra viewed under a Leeuwenhoek replica microscope

As is well known, Leeuwenhoek made his own single lens microscopes for his studies. Single lenses were superior to compound microscopes of the time and remained so until the development of achromatic microscope objectives in the early 19th century. I have two commercial replicas, the one shown left and the second sold by the Museum Boerhaave. Both have similar magnifications but I prefer the former for use as it's somewhat larger and easier to handle.

Left. A modern brass replica made by Chris Kirby of Christopher Allen Replicas (UK) with simply engineered parts and aged to look authentic in a style typical of Leeuwenhoek's designs

It is shown from the subject's side and has to be held close to the eye from the other side. The single lens is fixed between two riveted brass plates with dimpled apertures. The three screws allow focussing and subject orientation. The subject is mounted on the pin.

This replica is stated to have a ca. 100X magnification (at 250 mm) and was confirmed by my own measurements. The focal length is ca. 2.5 mm and subject field of view presented to the eye is ca. 0.8 mm.

The replica mounted on a Sony NEX 5N digital camera body. There are no optical components other than the replica lens. The assembly was mounted on a tripod and pointed at either a curtained window (to control light aperture) or an indoors lamp. Note that the adaptor was one to hand, but any lensless mechanical extension suitable for the camera mount can be used e.g. one or more extension tubes. The amount of extension will change the field captured which could be chosen as the full circular field of the lens or a greater extension to crop the field to suit.

The lens to sensor distance was 63 mm i.e. much less than the traditional 250 mm projection distance for formal studies. This projection distance was more practical for the setup and image just filled the APS sensor. A magnification more typical of what is seen by the eye is restored in either a screen or printed image. (See below).

In letters earlier than Sept. 7th 1674, Leeuwenhoek described his use of fine capillary tubes for studying liquids such as blood or milk and may have used such tubes for his aquatic samples from Berkelse Lake. As Wim notes in his article, if he

was reporting *Spirogyra* using such tubes it must have been a challenge persuading the long filaments to enter a narrow tube compared to the more likely candidate which Wim suggests.

Leeuwenhoek is known to have also used mica plates or thin blown glass to mount aqueous subjects attached to the pin (1) and I adopted a similar method as shown using coverslip pieces but with a more suitable support. Hans Loncke shows modified pin designs for his work with the splendid Leeuwenhoek replica that he built (2). Perhaps Leeuwenhoek also made alternative supports to the pin for work with flat plates.

Below. *Spirogyra* viewed under a Leeuwenhoek replica microscope. Optical mag 100X. Typical filament diameters 36 μm.

A darkened room with a vertical narrow light source was used to mimic Leeuwenhoek's suggestion for best use of his microscopes i.e. with a restricted aperture (3,4).

The Christopher Allen replica uses a hand ground 'convex' glass lens by the maker Chris Kirby (5). Leeuwenhoek was known to have made and used ground lenses or blown aspheric lenses (6).

A The lens shows the cellular structure of the filament clearly and the spiral chloroplasts and pyrenoids. Residual aberrations aside, the image differs not that much from a modern achromatic objective on a compound microscope, see later section.

B The same filament as above except using darkfield. Leeuwenhoek could not have failed to create and value the use of darkfield illumination with appropriate subjects. When setting up the microscope with a narrow light source, until the lens is fully aligned a slightly off-axis view readily creates this form of lighting.

For a discussion of Leeuwenhoek's likely use of darkfield see Snyder (4) and Dobell (7).

C The upper filament may be in the early stages of conjugation (or a different species!).

D A more general view of multiple filaments with ca. 50% of the full visual field is shown. The visual field of view of this 100X lens is ca. 0.8 mm which is comparable to that of a 16X modern compound microscope objective with 10X eyepiece (field. no. 18) and Optovar set at 1.25X on my Zeiss Photomicroscope III.

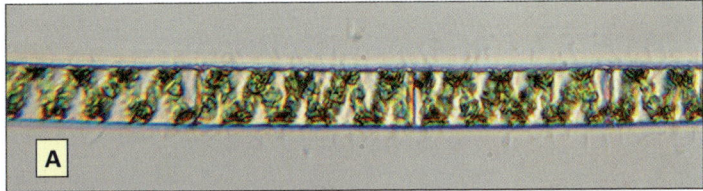

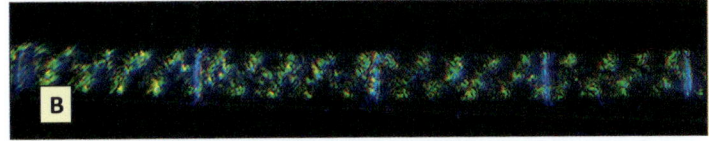

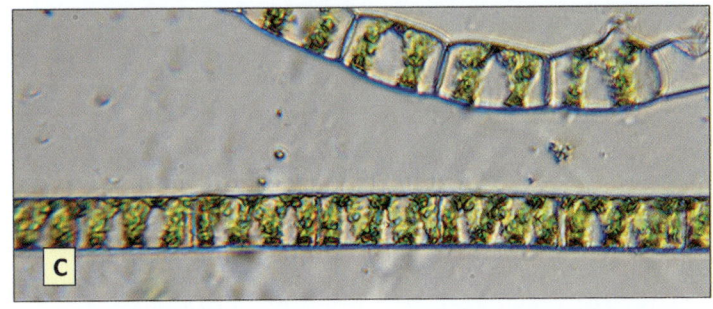

School studies of *Spirogyra* in the early 70s

My first encounter with *Spirogyra* was in the early 70s when studying for the GCE 'O' level exam in biology at high school. At that time, the course included studies of typical examples of various groups. Amoeba was the single-celled animal, *Spirogyra* the algae and advancing to more complex organisms like the hydra. A browse through some equivalent exam level textbooks using Amazon UK's 'Look Inside' feature suggests that this isn't now a typical modern approach in the UK.

They say that you never forget the schoolteachers who were of most influence and that is certainly the case for me. Mr Tan (or Tann?), the biology teacher at King Edmund Comprehensive, Rochford, Essex in the early 70s presented biology with an infectious enthusiasm and covered many aspects that still engage me as a hobbyist today. Practical microscopy work used the LOMO Biolam, a model that I later bought and have used as a hobbyist for over 35 years. Being a bit of a hoarder, I still have my school notes when aged 15 and the drawings are shown below

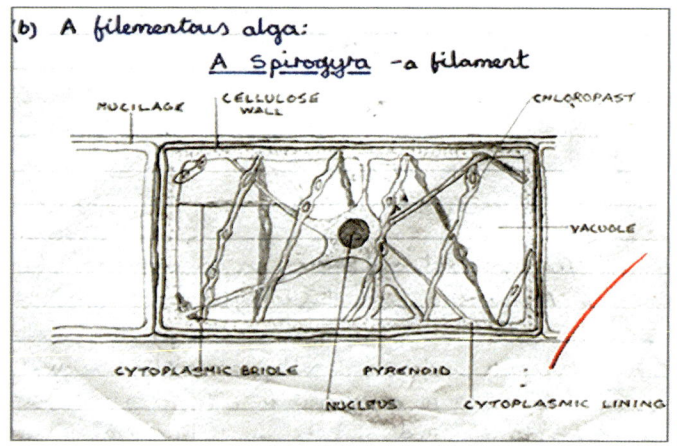

Above. Course work that likely required a redrawing of the diagram of an idealised cell from the accompanying textbook.

Left below. Own view of a specimen as seen under the LOMO Biolam microscope which were widely used in practicals.

Exploring *Spirogyra* using different lighting techniques including transmitted autofluorescence using a darkfield stop

The ease of preparing temporary fresh mounts of algae such as *Spirogyra* and the variety of forms it offers if undergoing conjugation, make it an interesting subject to study. Chlorophyll is noted for its relatively bright autofluorescence (cf. weakly emitting fluorochromes) and it is possible to use a normal transmitted compound microscope using a darkfield stop and a couple of filters to explore autofluorescence—no special epi-fluorescent microscope with intense lamp is required. Although a digital camera with good long exposure capabilities is ideal to record and study the results.

The following images used a Zeiss Photomicroscope III with Canon 600D DSLR body with Zeiss 10x Kpl eyepiece on short collar for projection. The 'D' setting on a water immersed achromatic-aplanatic condenser was used for both the darkfield and the autofluorescence studies.

The next three images show the same view of fresh *Spirogyra* in a temporary water mount under different lighting conditions.

Above. **Phase** *with Zeiss 10/0.22 achromatic phase objective. One filament shows the completion of scalariform conjugation where the contents of one filament (designated the 'male') are transferred to a second (the 'female') to form a zygote.*

Below. Same as previous:

(See opposite page for more)

Left. Spirogyra was a popular example of an algae, partly because of its distinctive forms of sexual reproduction. The scalariform of reproduction is shown.

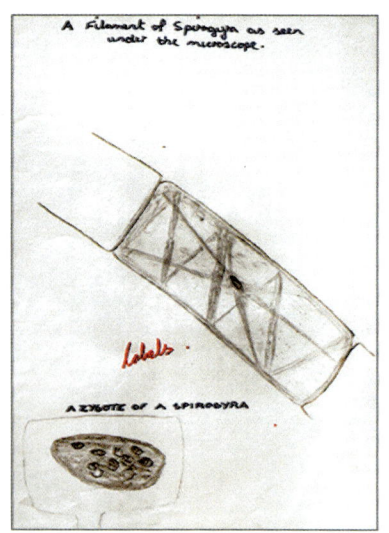

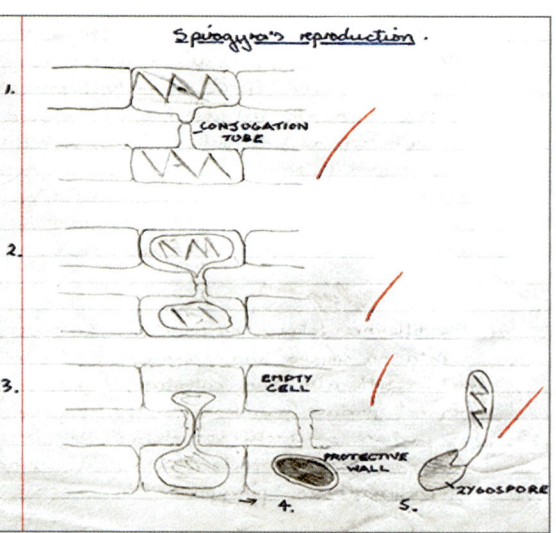

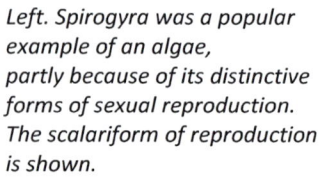

*10/0.22 objective with **darkfield** by using the larger Ph3 annulus (condenser water immersed).*

*Below. Zeiss 10/0.32 planapo objective with the Ph2 disc to create **circular oblique illumination** or COL (condenser water immersed).*

*Below. Zeiss Neofluar 25/0.65 objective with **DIC** and the type II prism. The shallow plane of focus provides a clearer view of the spiral chloroplasts and pyrenoids compared with the same view using phase below.*

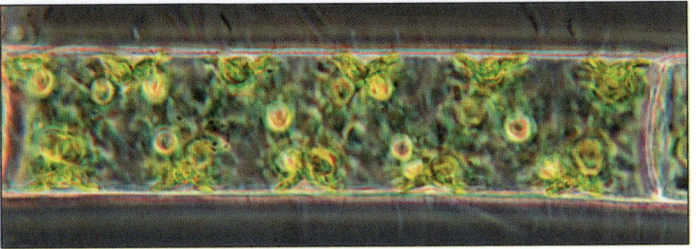

As above but using phase which gives a rather muddled view cf DIC.

Rather intriguingly, I was struggling to see either the nucleus or the cytoplasmic bridle, despite all the firepower the PMIII techniques offered. I had clearly drawn these features at school as shown earlier using brightfield on the LOMO. Whether they were indeed very clear for that species or it was wishful thinking knowing what the ideal cell in the textbook looked like, I'm not certain!

Autofluorescence of the chlorophyll containing chloroplasts shows the spiral structure well. (Below).

Zeiss 10/0.32 objective with darkfield using the water immersed 'D' setting on the condenser.

The standard 100W quartz halogen lamp at maximum intensity was used.

Exposure 30 secs ISO 800. Visual studies are possible with a standard 100W halogen lamp if the eyes are adapted in a dark room but photography is a better way to examine the effect.

The filters best used for chlorophyll autofluorescence from past experience were intentionally 'leaky' i.e. some blue excitation light was allowed to pass the barrier filter as it defines non-fluorescing components e.g. the cell walls in pale blue darkfield. Small e.g. 18 mm filters as used in the Zeiss III RS head are fine.

Excitation, 1 Schott BG12 filter (two would be usual for the deep blue excitation set in the III RS epi head). CM500S filter (or a BG38) to remove residual red from the light. Both filters sit on the field lens plate.

Barrier filter. Barrier '478 nm' filter from the III RS epi head. The barrier '500 nm' should be used for the Zeiss deep blue set but the 478 nm lets some blue pass. This filter sits in the PMIII filter holder supported on a card collar.

The following two images are of the same view to compare phase and autofluorescence. Two parallel filaments have completed the scalariform form of sexual reproduction. The zygotes are shown and the empty 'male' filament cells..

Below.
Same subject and field of view as above but autofluorescence. The chlorophyll rich zygotes show well in addition to the chloroplasts in the lower filament. Exposure 30 secs ISO 800.

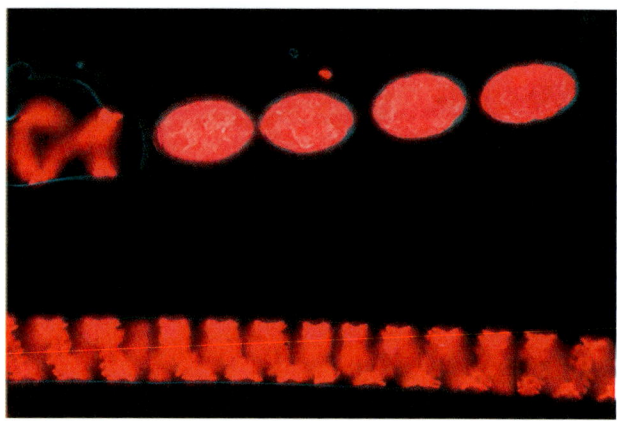

An attempt at identifying to species

Identification to species usually requires the reproductive forms to be present. As this was the case with the sample above collected from the water trough, I had a stab at an ID. I'd treated myself some years ago to the splendid *The Freshwater Algal Flora of the British Isles* edited by John, Whitton and Brook pub. 2002 (shown right with dish of *Spirogyra*). This was a more affordable two figure sum at the time but the second edition pub. 2011 is now typically £140. The flora includes a key, supported with many illustrations, to the 50 or so species of the 400 in the *Spirogyra* genus which they note have been reported in the British Isles.

Excellent book

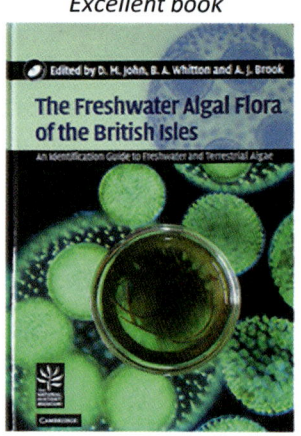

For the dominant species present, the typical measurements were: filament diameter 36 µm, cell length 68 - 130 µm, ellisoidal zygotes 58 µm x 30 µm Chloroplasts 1 per cell? If I've keyed out correctly (having never used the key before) it is *Spirogyra varians* (Hassall) Kützing 1849. Noted as 'probably cosmopolitan'. Mountain streams are included as a habitat which the slow flow through the upland water trough it was collected from could be regarded as. Although a feature noted on the zygotes for this species is 'mostly with a distinct suture line' which I could not see.

Some commercially prepared slides

The late and sadly missed Eric Marson of Northern Biological Supplies (NBS) offered a splendid set of fluid mount spirogyra slides. They included examples of both the scalariform and lateral forms of sexual reproduction. The former is the commoner 'ladder' form where two filaments come side by side, the lateral form is where adjacent cells in the same filament undergo reproduction. Sadly, only one of the fluid mounts is still in good condition, the others have dried out and are now unusable. I believe that one or more of these samples also formed the basis of Eric's paper which he published in the *Quekett Journal* (8) on his meticulous observations on aspects of selected *Spirogyra* species and their reproduction.

Many of the good value large slide sets contain examples of spirogyra. My brother Ian and I have an unbranded 100 slide set with two examples as shown. Unfortunately, this may be an example of false economy rather than selecting slides of particular interest from a well established preparer.

The quality of the mounts are variable and the mount has crystallised in many slides, making any photography using contrast enhancement unsuited as shown. Note also the curious labelling 'Spriogyra Conjugation' and 'Spirogation'.

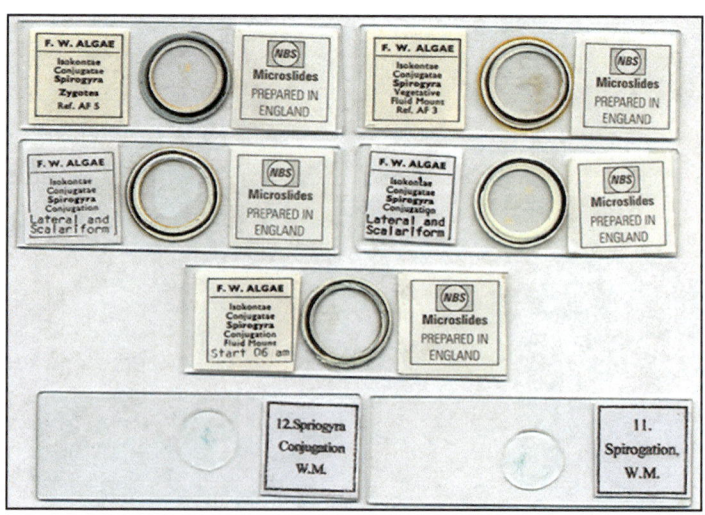

NBS Slides

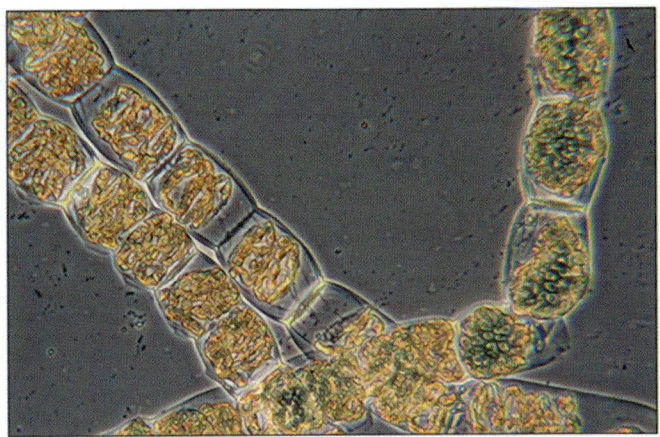 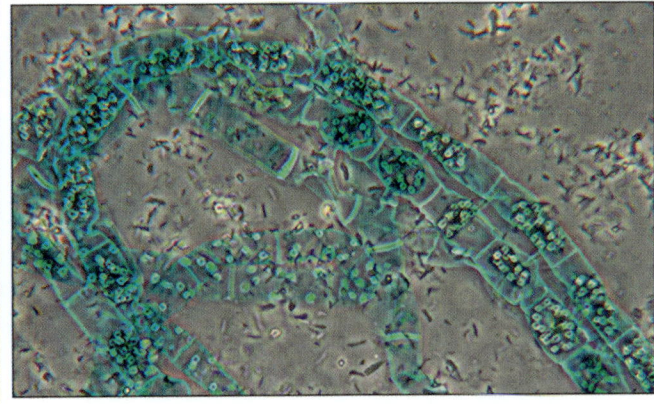

Left, filaments from the 'Spirogyra, Conjugation, Fluid Mount, Start 06 am' slide by NBS above. The slide remaining fluid from the set. Right, filament from the unbranded slide 12 shown above. The mount has crystallised giving muddled views if contrast enhancement is used. The filaments looks as if they have been stained.
Both slides using a Zeiss 10/0.22 achromatic phase objective in phase.

The author David Walker welcomes any comments / corrections. Email: micscape@ntlworld.com

References

1. M. Folkes, 'Some account of Mr. Leeuwenhoek's curious Microscopes, lately presented to the Royal Society', *Philosophical Transactions*, 1724, XXXII, 446. As quoted by Dobell ref. 7, p. 316. (Link is to open access paper.) *http://rstl.royalsocietypublishing.org/content/32/370-380/446.full.pdf+html*

2. e.g. Hans Loncke in his *Micscape* July 2007 article 'Making an Antoni van Leeuwenhoek microscope replica'. *www.microscopy-uk.org.uk/mag/artjul07/hl-loncke2.html*

3. Brian J. Ford, 'The Leeuwenhoek Legacy', 1991, Biopress, London, p.73 side note who cites ref. 3b.
3b. Antoni van Leeuwenhoek in his letter dated June 1st 1674 to H. Oldenburg Secretary of the Royal Society, *The Collected Letters of Antoni van Leeuwenhoek = Alle de Brieven van Antoni van Leeuwenhoek*, 1939, vol. I, letter no. 8 [4], p.115. (Link is to open access letter.) *www.dbnl.org/tekst/leeu027alle01_01/leeu027alle01_01_0010.php#b0008*

4. Laura J. Snyder, 'Eye of the Beholder. Johannes Vermeer, Antoni van Leeuwenhoek, and the Reinvention of Seeing', pub. Head of Zeuss Ltd., 2015, p.294-5 who cites ref. 4b.
4b. Barnett Cohen, 'On Leeuwenhoek's Method of Seeing Bacteria', J. of Bacteriology, 1937, 34(3), 343-346. (Link is to open access paper.) *http://www.ncbi.nlm.nih.gov/pmc/articles/PMC545235/*

5. Phil Greaves, 'Replica Leeuwenhoek microscopes', on the Quekett Microscopical Club website accessed February 2016. *www.quekett.org/resources/article-archive/replica-leeuwenhoek-microscopes*

6. J. van Zuylen, 'The Microscopes of Antoni van Leeuwenhoek', *J. of Microscopy*, 1981, 121/3, pp. 309-328. *(Reprinted in 'Antoni van Leeuwenhoek, 1632-1723', Eds. L. C. Palm and H.A.M. Snelders, Rodopi, Amsterdam, 1982, pp.29-55.)*

7. C. Dobell, 'Antony van Leeuwenhoek and his "Little Animals"', first published 1932. Dover Edition 1960, p.331. (Link is to a free copy on www.archive.org.) *archive.org/details/antonyvanleeuwen00dobe*

8. J. E. Marson, 'Observations on Sexual Reproduction in species of *Spirogyra*', *Microscopy (The Journal of the Quekett Microscopical Club* now the *Quekett Journal of Microscopy)*, 1992, vol. 36 (part 9), 712-717, 720. *www.quekett.org*

Acknowledgement
Thank you to the *Digitale Bibliotheek voor de Nederlandse Letteren (DBNL)* website for both hosting and making freely accessible the first fifteen volumes of the 'Collected Letters of Antoni van Leeuwenhoek'. *www.dbnl.org/auteurs/auteur.php?id=leeu027*

Related *Micscape* articles
'*Spirogyra*' by Jan Parmentier, *Micscape* January 1999. *www.microscopy-uk.org.uk/mag/artjan99/gyra.html*

'Conjugation in *Spirogyra*', by Wim van Egmond, *Micscape* 1998. Part of Wim's *The Smallest Page on the Web* suite. *www.microscopy-uk.org.uk/mag/wimsmall/spirogyraconju.html*

'Forays into fluorescence. Simple transmitted blue light autofluorescence of mosses and algae imaged with a digital SLR.' by David Walker, *Micscape* March 2009. *www.microscopy-uk.org.uk//mag/artmar09/dw-fluor1.html*

N1
www.microscopy-uk.org.uk/mag/artfeb16/wimleeuwenhoek2.html

Revision history.
First published February 13th 2016.
Various amendments, corrections and additional references added up to and including February 20th 2016.
Format change for printed version: July 2016. Web address: www.microscopy-uk.org.uk/mag/artfeb16/dw-spirogyra.html

The riddle of the 'green streaks'. In search of the first microorganism which Antoni van Leeuwenhoek described.

by Wim van Egmond, the Netherlands
in collaboration with Frans Kouwets
Published February 2016

Portrait of Van Leeuwenhoek by Johannes Verkolje.

During the summer of 1674 Antoni van Leeuwenhoek (1632-1723), a draper from Delft, filled a small bottle with water from a lake called the Berkelse Meer. He examined the contents with his microscope the next day. His description of the find and what he observed using his tiny but superb microscope can be regarded as the beginning of microbiology. He wrote about this discovery in Dutch. His letter, dated September 7th, was sent to the Royal Society in London, translated into English and an extract published in *Philosophical Transactions.* Although the language in which he wrote is a bit old-fashioned it is still a treat to read. Here is the English translation. (The original letter was one long paragraph, I split it into 3 parts for easier reading.)

"About two hours distant from this Town there lies an inland lake, called the Berkelse Mere, whose bottom in many places is very marshy, or boggy. Its water is in winter very clear, but at the beginning or in the middle of summer it becomes whitish, and there are then little green clouds floating through it ; which, according to the saying of the country folk dwelling thereabout, is caused by the dew, which happens to fall at that time, and which they call honey-dew. This water is abounding in fish, which is very good and savoury.

Passing just lately over this lake, at a time when the wind blew pretty hard, and seeing the water as above described, I took up a little of it in a glass phial ; and

examining this water next day, I found floating therein divers earthy particles, and some green streaks, spirally wound serpent-wise, and orderly arranged, after the manner of the copper or tin worms, which distillers use to cool their liquors as they distil over. The whole circumference of each of these streaks was about the thickness of a hair of one's head. Other particles had but the beginning of the foresaid streak ; but all consisted of very small green globules joined together: and there were very many small green globules as well.

Among these there were, besides, very many little animalcules, whereof some were roundish, while others, a bit bigger, consisted of an oval. On these last I saw two little legs near the head, and two little fins at the hindmost end of the body. Others were somewhat longer than an oval, and these were very slow a-moving, and few in number. These animalcules had divers colours, some being whitish and transparent ; others with green and very glittering little scales ; others again were green in the middle, and before and behind white ; others yet were ashen grey. And the motion of most of these animalcules in the water was so swift, and so

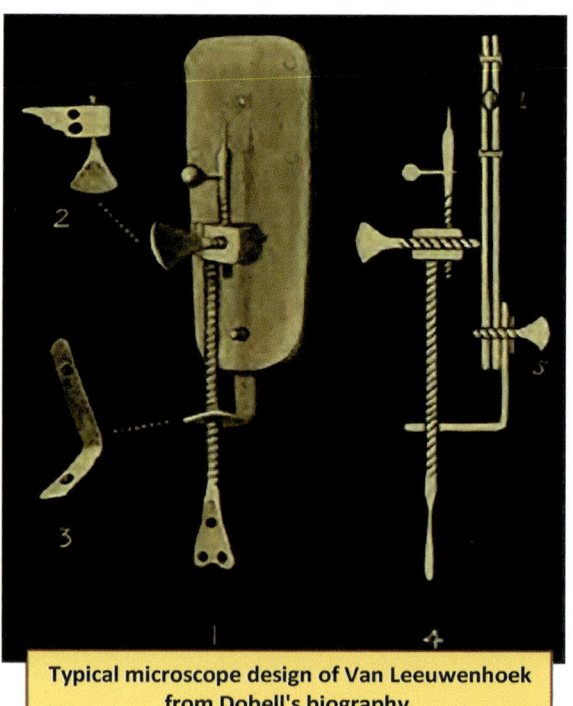

Typical microscope design of Van Leeuwenhoek from Dobell's biography.

various, upwards, downwards, and round about, that 'twas wonderful to see : and I judge that some of these little creatures were above a thousand times smaller than the smallest ones I have ever yet seen, upon the rind of cheese, in wheaten flour, mould, and the like."

There has always been a lot of speculation about the true nature of the organisms he described in this letter. Van Leeuwenhoek was a layman and it was long before Linnaeus provided us with a systematic way of describing living organisms. Almost everything Van Leeuwenhoek observed through his simple single lens microscope had never been seen before. Van Leeuwenhoek did not give the organisms names, he simply described their appearance, as detailed as he could.

He was one of the best observers of his time. In any case literally because he was the only one who could make a microscope with a magnification of up to several hundred times and with a very high resolving power.

The Berkelse Meres were situated south east of Delft, where Van Leeuwenhoek lived.

Spirogyra photographed in darkfield illumination. Image width: approximately 0,9 mm.

One of the most influential scientists who studied the work of Van Leeuwenhoek was Clifford Dobell (1886-1949). He was a protozoologist who studied intestinal amoebae but also published on algae. His classic biography about Van Leeuwenhoek from 1932 Antony van Leeuwenhoek and his little Animals[1] from which the above translation was sourced was very popular and most of the names put to the descriptions of Van Leeuwenhoek can be credited to Dobell. He thought the spirally arranged green 'streaks' Van Leeuwenhoek described were Spirogyra, a filamentous alga that has coiled chloroplasts. Ever since, this is regarded as the organism Van Leeuwenhoek described.

Dobell comments (Footnote 2, p. 110): "The common green alga *Spirogyra*: the earliest recorded observations on this organism. The size of the filament negatives the suggestion that L. could have been referring to *Arthrospira* or *Spirulina*, (these are filamentous cyanobacteria)."

He furthermore gives suggestions for the other organisms Van Leeuwenhoek mentioned: ciliates, rotifers and *Euglena*. We can be pretty sure he observed at least one of these groups in the sample, but the description of these organisms is not as detailed as the first one. The present article is about that first observation, the most important one, the first observation of a living aquatic microorganism. Dobell dismisses *Arthrospira* and *Spirulina* because of their size. I agree. But could there be other candidates?

In 2009 our Dutch desmid club (www.desmids.nl) organised an excursion to the Czech Republic to collect some samples of this group of beautifully shaped green algae. During a field trip I was talking with Frans Kouwets, phycologist at the Ministry of Infrastructure and Environment in Lelystad. Somewhere during our conversation I mentioned *Spirogyra* as the organism found by Antoni van Leeuwenhoek. 'Aha', Frans said, 'that is something that I seriously doubt. I don't think it was *Spirogyra* and perhaps it is interesting to publish something about this.'

It is now 2016 and Frans and I have discussed this subject over the years. I have studied Van Leeuwenhoek for my work for a museum about microbes, continuously finding new clues. An extra reason why I am so interested is the fact that I grew up near the location and live near the former Berkelse Meer. Frans currently is very busy with his monograph of the desmid genus *Cosmarium* and encouraged me to write the article. But Frans is reading over my shoulder so don't forget that he started all this.

Let's start to sum up the reasons why we think *Spirogyra* is an unlikely candidate. And perhaps it is a good moment to refer to the motto of the Royal Society; *Nullius in verba*, which can be translated as 'take nobody's word for it'.

Spirogyra is common in shallow ponds and ditches. A large lake as the Berkelse Meer was (1.5 kilometer wide) is not the usual habitat. Although it inhabits lake edges it is not planktonic.

Spirogyra does form floating masses but in early summer, and then their colour is not whitish green but yellow or brown green.

Spirogyra blooms consist of very long filaments or strands, not microscopic streaks. Being such a keen observer, Van Leeuwenhoek was always very accurate in his observations. He would certainly have described *Spirogyra* as long strands. If you find *Spirogyra* you can take it out of the water with your hands as scum, and it feels slippery.

Van Leeuwenhoek used capillary tubes at the time for studying liquids e.g. in his earlier letter dated July 6th 1674[2]. If he did it would be impossible to put *Spirogyra* in such a tube.

Spirogyra bloom, forming floating algal beds, photographed in June.

There is enough reason for doubting it was *Spirogyra.* It is interesting to see that for such a long time everybody copied Dobell's suggestion, although it is understandable. What else could it have been? Before we search for other candidates, let's look again at the description.

"I found floating therein divers earthy particles, and some green streaks, spirally wound serpent-wise, and orderly arranged, after the manner of the copper or tin worms, which distillers use to cool their liquors as they distil over. The whole circumference of each of these streaks was about the thickness of a hair of one's head. Other particles had but the beginning of the foresaid streak; but all consisted of very small green globules joined together : and there were very many small green globules as well."

And the original letter in Dutch:

"bevonde ick daer in te drijven, verscheijde aertsche deeltgens, ende eenige groene ranckjens, in geschickte ordre slanghs gewijse omgekrult, op gelijcke manier, als de copere off tinne slangen sijn, die de distelatuers gebruijcken, omme haer over gehaelde wateren te verkoelen, ende de gantsche circumferentie, van jder van dese ranckjens, hadt ontrent de dickte van een haer van ons hooft; andere deeltgens hadden maer een begin, van het boven verhaelde ranckje, alle bestaende uijt seer kleijne groene same gevoeghde clootgens, als mede seer veel kleijne groene clootgens,"

Van Leeuwenhoek only wrote in Dutch. He did not know any other language and his Dutch is translated by others. That is why I think we should look at the original Dutch text and try to read it as carefully as possible. The actual Dutch word Van Leeuwenhoek used was 'ranckje' (plural ranckjens). While writing this article I noticed that the English translation of the word 'ranckjes' (both in the *Collected Letter's* and Dobell's), into 'streaks' is not accurate. The word streak is a bit misleading because it suggests an elongated shape. In my opinion a correct translation would be small tendril. And I am sure he used the Dutch equivalent because a tendril is often coiled. The Dutch text is even clearer on the coils than the translation.

It can be translated various ways. Here's what I would make of it: Several earthy particles and some green tendrils that are bent snake-wise to orderly arranged curls, like the copper or tin tubes distillers use to cool their transferred water.

Next: why would Van Leeuwenhoek mention 'the whole circumference' for the thickness of *Spirogyra*? If you look at *Spirogyra* it is not necessary to add 'the

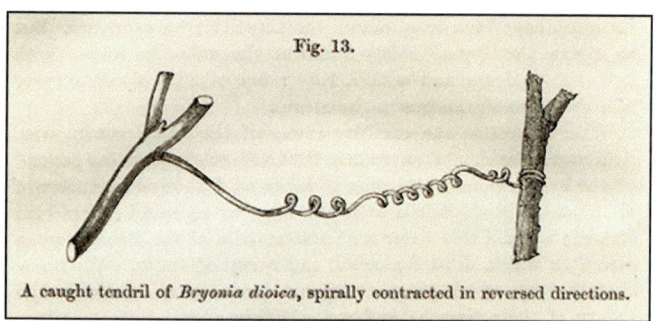

Illustration from Darwin's book on the
Movements and Habits of Climbing Plants - 1854

whole circumference'. It should be enough to describe it as 'long strands the thickness of a human hair'. He only had a reason to mention 'the whole circumference' when it is not obvious whether that thickness refers to the coil or what the coil is made of. Could it be that he wrote 'the whole circumference' because the organism he described actually was a coil and not a tube with coils inside such as *Spirogyra*?

Finally, the last part of the sentence was translated

A distiller's copper worm

as, "Other particles had but the beginning of the foresaid streak; but all consisted of very small green globules joined together : and there were very many small green globules as well." Apart from the streaks the translation is good; clootgens are little balls. Other particles had only a beginning of the small tendrils mentioned above, all made out of very small green little balls joined together, as well as many small green balls. [This last double remark is a bit odd but I think he means there were also separate little balls]. (Note: *Spirogyra* can be found as very small cells but the word particle would not be a good description. It is also questionable if these small cells of *Spirogyra* can be found floating.)

Van Leeuwenhoek literally describes his find as regularly ordered coils made out of little balls (clootgens). Resembling a vine tendril and a distiller's copper worm (shown above). Many of his descriptions are cryptic and hard to associate with certain organisms. That is true for the other organisms he describes in the latter part of the letter. But this section, the first time aquatic microbes are ever described, is clear and accurate. We also have the details of the time and setting. In his boat on the Berkelse Meer, a nutrient rich lake (plenty of fish) Antoni van Leeuwenhoek. describes

Below: Cyanobacterial bloom of *Dolichospermum*, photographed in darkfield illumination. Image width: approximately 0,9 mm

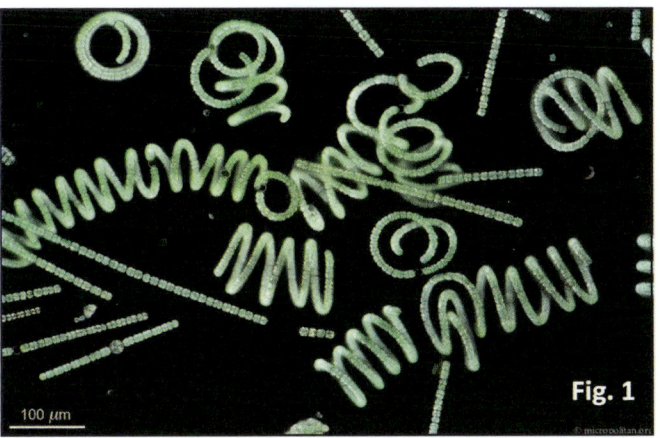

100 μm

Fig. 1

a late summer cyanobacterial bloom and the first aquatic microbe captured in words could very well be the one shown here: (*Fig. 1*)

The image shows what Frans and I suggest to be the most likely candidate. It's a filament-forming cyanobacterium. They are made out of little balls, the individual cells, that form very regularly coiled strands. I found these myself in late summer in a large lake as a whitish green layer against the surface. During August there was a bloom of these cyanobacteria. So the time and location also fits the description. I have collected them the same way Van Leeuwenhoek did. They float under the surface as a cloudy layer and when you put a bottle in the water you can easily fill it.

Cyanobacteria, or blue green algae, are bacteria that perform photosynthesis. Although they have a bad reputation they are important oxygen producers. The cyanobacterium that we think Van Leeuwenhoek described is in the genus *Dolichospermum*. Recently the planktic forms of *Anabaena* were separated from this genus and transferred to the genus *Dolichospermum*. The coils of this species of *Dolichospermum* have a diameter that range between 70 and 90 microns. Van Leeuwenhoek is talking about the whole circumference (not the little cells) measuring about the same as the thickness of a hair from our head. According to the DNBL Collected Letters[3] website the size of such a hair is between 60 and 80 microns.

Dobell ruled out the helical cyanobacteria *Arthrospira* and *Spirulina* and I think he was right. Their filaments are more slender but more importantly, they are not made out of little balls.

Frans and I can't think of any other suitable candidate. If we compare all the arguments in favor of *Dolichospermum* and all the arguments against *Spirogyra* there is only one conclusion. It is much more likely that Van Leeuwenhoek observed cyanobacteria like Dolichospermum. And that would mean it is not only the first account of living aquatic microorganisms, but also the first observation of bacteria, two years before Van Leeuwenhoek described them in his studies of 'pepper water' in the letter dated October 9th 1676[4].

I think it is important to state that we can only speculate here. We can't be 100% sure. But in the case of the greenish helix of little balls we can safely replace Spirogyra by *Dolichospermum*, the cyanobacterium formerly known as *Anabaena*, as the number one candidate of the first described aquatic microorganism.

Unless...

What about the 'earthy particles'? If we assume that the whitish water with the green clouds is a cyanobacterial bloom, then the earthy particles could represent cyanobacteria as well. Such a bloom often consists of several genera of cyanobacteria. The earthy particles could be *Microcystis*. Their colonies have shapes that can be described as earthy particles.

And then that would be the first aquatic microorganism ever described.

But I think here a *Nullius in verba* would do.

Footnotes

There are many different species of *Dolichospermum* with various dimensions that form coiled filaments so it could well be that Van Leeuwenhoek saw a different species. But there are not that many species with a regular coil of the diameter Van Leeuwenhoek mentions. If we follow the description of a regular coil with the diameter between 60 and 80 micron we can narrow it down to *D. circinale*, *D. spiroides* or *D. crassum*. The one I photographed may be *D. crassum* but not without doubt because these species vary in appearance and are hard to distinguish from one another.

A paper by Jiří Komárek and Eliška Zapomelová published in *Fottea* (Journal of the Czech Phycological Society), 2007, vol. 7, issue 1, 1-31, presents a detailed and richly illustrated review of these coiled cyanobacteria. Planktic morphospecies of the cyanobacterial genus *Anabaena* subg. *Dolichospermum* – 1. part: coiled types[5].

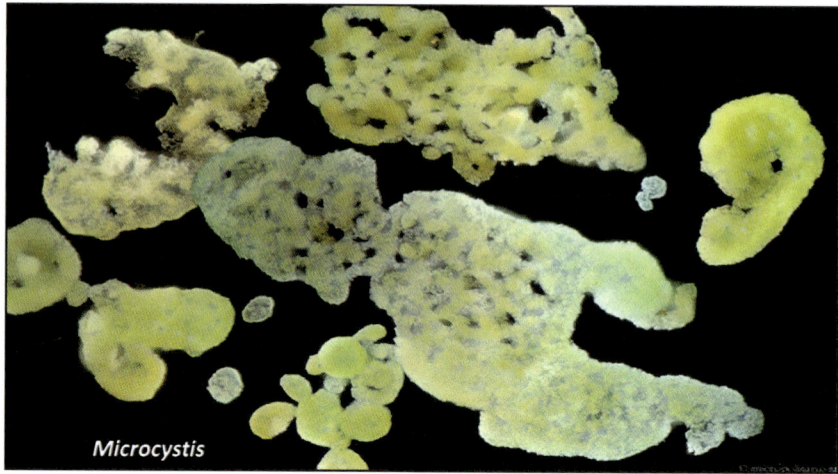

Microcystis

Acknowledgements

This article was written in collaboration. I would like to thank Frans Kouwets for allowing me to publish this investigation that started with his initial idea and for the feedback during the writing of this article.

Micscape's editor David Walker was a great help and offered valuable suggestions. Read his article 'Reflections on studying Spirogyra[6]' which includes images of a Spirogyra species taken using a commercial replica of a Van Leeuwenhoek microscope. *(See p.32 of this Yearbook).*

I also would like to thank Richard Howey and Maria van Herk for their feedback and help.

Thank you to the Digitale Bibliotheek voor de Nederlandse Letteren (DBNL)[7] website for both hosting and making freely accessible the first fifteen volumes of the 'Collected Letters of Antoni van Leeuwenhoek'.

Image credits

All material (except the maps, tendril, copper tin image and images as described below) © Wim van Egmond.

The portrait by Johannes Verkolje of Antony van Leeuwenhoek F.R.S. from the frontispiece of Hoole's 'The Select Works of Antony van Leeuwenhoek', published in two vols. ca. 1800. Now in the public domain at www.archive.org.

Van Leeuwenhoek diagram from Plate XXXI of Clifford Dobell's 'Antony van Leeuwenhoek and his "Little Animals"' in the public domain at www.archive.org.

The Berkelse Meres Map is an enhanced detail from: 't Hooge heemraedschap van Delflant / Nicolaes en Jacob Kruikius (1712). Source: Hoogheemraadschap van Delfland te Delft inv.nr. OAS 726 from the Delft University of Technology website[8] with thanks.

The distiller's copper worm image is web sourced but exact credit uncertain, but will be pleased to add due credit if contacted. Comments to the author Wim van Egmond (*t936927@telfort.nl*) are welcomed.

Visit the Micropolitan Museum[9].

References:

1)archive.org/details/antonyvanleeuwen00dobe

2)www.dbnl.org/tekst/ leeu027alle01_01leeu027alle01_01_0011.php#b0009

3)www.dbnl.org/tekst/leeu027alle09_01/ leeu027alle09_01_0025.php

4)www.dbnl.org/tekst/leeu027alle02_01/ leeu027alle02_01_0006.php#b0026

5)www.fottea.czechphycology.cz/artkey/fot-200701-0001_Planktic_morphospecies_of_the_cyanobacterial_ genus_Anabaena_subg_Dolichospermum_-_1_part_coiled_types.php

6)www.microscopy-uk.org.uk/mag/artfeb16/dw-spirogyra.html

7)www.dbnl.org/auteurs/auteur.php?id=leeu027

8)tresor.tudelft.nl/kaarten/webpages/KVD2001_03.html

9)www.microscopy-uk.org.uk/micropolitan/index.html

THE 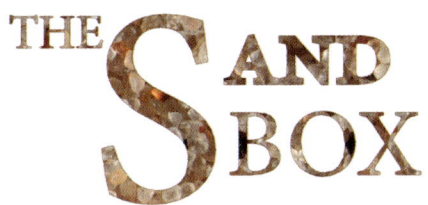 AND BOX

by Kirsten Martin
Published December 2015

Sand is a material that is fairly common, yet still draws the attraction of many collectors around the world. Not only do these individuals collect the sand themselves; they barter and trade for it with other collectors. Becoming a psammophile, or sand collector, is as easy as scooping a little sand from a nearby beach into a plastic container and carrying it home. So what is it about this grainy matter that makes it so special to so many people? Many people will cite a feeling of wonder when looking closely at the differences from one sample of sand to the next. Sand contains different minerals that can cause it to be different colours based on the origin of the grains. It also can be found almost anywhere so there is never a shortage of sand to observe. It's very important in nature, as well as in industrial items. Part of the fun of sand collecting is also looking around at where the sand was found so as to obtain a better understanding of how the sand got there.

Sand itself is just finely divided rock and is made up of small particles or granules called sand grains. It is commonly transported by wind or water and can appear in the form of beaches, dunes, sand spits, and sand bars. It can even make up the majority of an area's soil composition, as it does in deserts. In more specific, geological terms sand is made up of particles ranging in size from about one-sixteenth of a millimetre to two millimetres in diameter.

Anything smaller than this is considered a different category called silt. Anything larger is also considered a different category called gravel. One way to test if the material you have found is silt or

sand is to rub it between your fingers. If it feels grainy it is sand. If it feels smooth like flour it is silt. The size and shape of sand grains is largely determined by the area in which the sand is found or originated from. Local rock sources are the source of the

Top: Picture Rocks Pack, MI reflected light
Bottom: Picture Rocks Pack, MI cross polarized light

sand, and the way those rocks break down and become weathered determines what the sand will look like. For instance, areas with primarily limestone will yield bright

Kirsten Martin is a Biomedical Photographic communications Major at Rochester Institute of Technology, with an immersion in Japanese.

Originally from Pennsylvania, she would like to travel to Japan and spend a few years working there after she graduates. Kirsten's love of science developed thanks to her father who never missed an opportunity to enrich her education outside of the classroom. When she later discovered her love of photography in high school, she was ecstatic to learn that science and photography could work together to create wonderful images. Kirsten is expecting to graduate in the spring of 2017 and is excited to see what the future holds for her.

white sands such as those seen in tropical beaches. Weathering and erosion of granite results in sand with a high feldspar content called arkose. Sands can also be various shades of red, green, gray or black, and brown. Sometimes, small gemstones can be found in sand samples.

Not all sand is actually crushed up rocks, though. Some sand, especially on beaches, can be made partially of marine life remains. Beaches can therefore tell scientists something about the diversity, abundance, and ecology of organisms that dwell in the nearby water. Sand with a carbonate composition indicates that those grains of sand were likely from biological origins. An easy way to test if sand has calcium carbonate in its composition is to remove a small portion of the sample and place a few drops of vinegar on it. This will result in the calcium carbonate dissolving in the vinegar and producing small bubbles of carbon dioxide. If the sand produces bubbles in reaction to the vinegar, it may be biological in origin. If the sand sample in question was taken from a desert, in combination with shell fragments, it could indicate that there may have been an ancient sea in or near that location. Sand can also be used to determine changing sea levels.

Sand can be found in a multitude of places, including the beach and deserts. It helps to shape the landscape of the places where it is found. On beaches the sand dunes occur where some vegetation is present to trap sand grains as they are carried by the wind. As the old vegetation dies, it provides nutrients for a new generation of vegetation to grow on the piles of sand. This traps more sand as the wind carries it and a dune is formed. The dune will continue to grow and sustain itself in this manner. Once a dune is established, the network of roots helps to provide a solid protection in the event of a minor storm. There are also man-made dunes, which are constructed using a bulldozer to push the sand into a pile near buildings as a measure for storm protection.

These dunes look very different from natural dunes, as they contain much larger particles than would naturally be carried to a dune by the wind. A few days after construction, most of the lighter particles of a man-made dune will be blown away and a layer of shells will be exposed – unable to be blown away by the wind and artificially placed there by a bulldozer. The artificial dunes are much more easily eroded by inclement weather and are usually washed away by the next storm.

Ontario, NY reflected light

Ontario, NY cross polarized light

Prince Edward Island reflected light

a real threat to people, buildings, and agriculture in some areas of Africa, the Middle East, and China. Sometimes, to stop the movement of the dunes, oil is poured over top of them. While this does halt the movement of the dunes, it is degrading to the environment. Some have suggested making sand fences, but the best possible design for such fences has not been agreed upon yet. Sand dunes can also experience sand avalanches, in which sand falls quickly along the steep side of the dune. These can be deadly if a person

The dunes in deserts are made in a similar way to the dunes found naturally on beaches. As sand grains are picked up by the wind, they are deposited wherever there is a "wind shadow" in which the air consistently slows down enough to drop the sand grains. In deserts, however, dunes have much more freedom to move around. Through a process called "saltation" the sand grains start to skip across the ground in a similar way to a rock skipping across the water. Sometimes, these sand grains crash into each other and start to form a sheet flow. When these sheet flows occur dunes can move tens of meters in a matter of minutes. This is

is caught in one.

Not all sand is found in nature, though. Some sands serve industrial purposes, such as in concrete. When concrete is mixed with sand, it forms cement. The sand in cement helps to fill pores and keep the concrete from cracking as it solidifies. Usually, the sand that is used in cement is silica.

This is generally found in the form of quartz. The sand helps to strengthen the concrete as well. Since silica is easily available and economical compared to other fine particles that could do a similar job it is the primary ingredient added to concrete along with gravel to make cement. Another use of sand is in art. When an artist wants a glass piece to appear "frosted" a fine silica sand is used in a sand blasting machine. The sand has a hardness that is higher than glass, so when the sand blaster turns on the sand particles are blown

Prince Edward Island cross polarized light

Above. Hamoa, Maui reflected light Below). Copper Harbor, MI reflected light

Oloololo, Maui reflected light

out through a hose at high speeds aimed at the area of the piece that the artist wants frosted. This causes the sand to make tiny scratches in the surface of the glass, which results in the frosted appearance. Once the machine is turned off, the artist must wait for the dust to settle before opening it and removing the piece. If breathed in repeatedly over time, the silica particles can cause a disease in the lungs known as silicosis. Silica sand is not only used in frosting glass, it is also a component when making many kinds of standard and specialty glass. The chemical purity of this glass is responsible for the colour, clarity, and strength of the glass that is made with it. It can be used in the production of everything from sheet glass for buildings, to glass food storage containers, and glass cups and other tableware. When it is pulverized, ground silica is used to make fibreglass insulation. This is all possible because industrial sand is high purity silica with closely controlled sizing.

Since sand has been such a pervasive element throughout many landscapes throughout the world, it is not surprising that it is also included in cultures and rituals around the world. In the Bible, sand

is often used to talk about quantities far greater than the human mind can grasp. In Genesis 22:17 God tells Abraham "I will surely bless you and make your descendants as numerous as the stars in the sky and the sand on the seashore." This is a more poetic way of saying that Abraham will have many many descendants. Both the Navajo people and the Tibetan Buddhist Monks use sand in different colors to create temporary artwork. The Navajo people use it in ceremonies to create sand paintings. Rituals and ceremonies are an important part of their culture,

used for everything from milestones in life, to asking for luck on a hunt, to common colds and illnesses. Almost all of these rituals and ceremonies involve sand paintings. The Tibetan Buddhists create beautiful sand mandalas with coloured sands. The sand mandalas take hours, even days to complete. They are meant to transmit positive energies to the environment and those who view them. Once they are completed, they are systematically wiped away as a reminder of the impermanence of life. The mixed sand is then distributed into flowing water to share the good energies into the world. In Japan sand takes on a slightly different significance. While it is also used in Japanese rock gardens, there is a special type of sand on the beach in Okinawa. Commonly known as Japanese star sand, people come to this beach to sort through the grains to find tiny spiked particles. Local legend states that these tiny star-like grains of sand are the skeletal remains of the children of the Northern Star and the Southern Cross, who were all killed by a giant serpent as they fell into the ocean. In actuality, these tiny marvels are the exoskeletons of protozoa that live on the ocean floor.

They wash onto the beach and are a somewhat popular candidate for scanning electron microscope (SEM) imaging. Sand has also played a role in Egyptian culture, though less directly. One of the contributing factors to the discovery of mummification was from the natural mummification process that occurs when a body is buried in sand in a warm, dry climate.

Sand has helped shape not only landscapes, but also culture, and human productivity. It is prolific and finds its way into almost every aspect of our lives in one way or another; even if it doesn't always look like sand. It is also an amazing indicator of what kind of surroundings it may have come from. Many people collect sand, from young children to older seniors. Every grain has a history to it. Some of it may even perform important functions in industrial uses. Sand is often taken for granted, but upon closer inspection there is much more than meets the eye.

Okinawa, Japan reflected light

Pen for size reference, reflected light

Local legend states that these tiny star-like grains of sand are the skeletal remains of the children of the Northern Star and the Southern Cross, who were all killed by a giant serpent as they fell into the ocean.

In order to take the pictures that were featured in this article, I used a Canon 5D Mark II and a 65mm macro lens. I set up a work station with a copy stand and an adjustable stage much like what one would see on a microscope. This allowed for minute movements without shaking the sand grains and therefore disturbing them. I used a shutter release cord to minimize any camera shake. I used a small piece of foam core to put the sand on and create a background. I used a single handheld flash to illuminate the sand in each shot. Each frame was taken at 5x, so a comparison of the images shows the sizes of the grains in relation to the grains in other samples. I also included an image of the tip of a ball point pen as a reference to give a more tangible idea of how small the grains of sand actually are. For the cross polarized images, I suspended the sand on a piece of glass between two polarisers and illuminated it from underneath.

Works Cited

"Analyzing the Origin of Different Sands." Analyzing the Origin of Different Sands. N.p., n.d. Web. 10 Dec. 2015. <http://www.msnucleus.org/membership/html/k-6/rc/rocks/5/rcr5_2a.html>.

BBC News. BBC, n.d. Web. 10 Dec. 2015. <http://www.bbc.co.uk/religion/religions/buddhism/customs/mandala.shtml>.

"Christian Wedding Sand Ceremony." People. N.p., n.d. Web. 10 Dec. 2015. <http://people.opposingviews.com/christian-wedding-sand-ceremony-2114.html>.

"Glass Making." Nisa. N.p., n.d. Web. 10 Dec. 2015. <http://www.sand.org/Glass-Making>.

"Grains of Sand." Grains of Sand. N.p., n.d. Web. 10 Dec. 2015. <https://www.classzone.com/books/earth_science/terc/content/investigations/es1607/es1607page01.cfm?chapter_no=investigation>.

"Hinduism." Google Books. N.p., n.d. Web. 10 Dec. 2015. <https://books.google.com/books?id=jID3TuoiOMQC&pg=PA276&lpg=PA276&dq=importance%2Bof%2Bsand%2Bin%2Bhindi&source=bl&ots=Cy4anwkrIW&sig=npP-BsnDEmhITM0cJzWGBZOS_YA&hl=en&sa=X&ved=0CGEQ6AEwCWoVChMIgtPL4aSbyQIVBjAaCh0G1gec#v=onepage&q=sand&f=false>.

"Navajo Sandpaintings." , Also Called Dry Paintings, Are Used in Navajo Curing Ceremonies. N.p., n.d. Web. 10 Dec. 2015. <http://navajopeople.org/navajo-sand-painting.htm>.

"New Views on Sand." New Views on Sand. Virginia Institute of Marine Science, n.d. Web. <http://web.vims.edu/bridge/sand.pdf?svr=www>.

"Sand." - New World Encyclopedia. N.p., n.d. Web. 10 Dec. 2015.

"Sand Analysis." Sand: From the Beach, River or Lumber Yard. N.p., n.d. Web. 10 Dec. 2015. <http://www.microscope-microscope.org/applications/sand/microscopic-sand.htm>.

"Sand Dunes." Coastal Care. Santa Aguila Foundation, n.d. Web. <http://coastalcare.org/educate/sand-dunes/>.

"Sand Minerals." Sandatlas. N.p., n.d. Web. 10 Dec. 2015. <http://www.sandatlas.org/sand-minerals/>.

"Sands - Arenology - 646Senesac." Sands - Arenology - 646Senesac. N.p., n.d. Web. 10 Dec. 2015.

"Star Sands: Okinawa's Incredibly Shaped Living Fossils." Star Sands: Okinawa's Incredibly Shaped Living Fossils. N.p., n.d. Web. 10 Dec. 2015. <http://scribol.com/science/star-sands-okinawas-incredibly-shaped-living-fossils>.

The Diverse Ciliate Community

By: Jason Dinelli

by Jason Dinelli
Published Dec 2015

Overview

There are millions of different microorganisms that can be found and identified in fresh water, but they mainly fall under two different kingdoms; the Bacteria Kingdom and the Protista Kingdom.

Bacteria are formed from only one cell, and are referred to as unicellular organisms, or prokaryotes.

While bacteria do have genetic material, it is suspended freely in cytoplasm and can easily move around inside the bacteria. There are two Kingdoms of bacteria: Eubacteria and Archaebacteria. Archaebacteria typically live in very harsh conditions that range from being very hot, to very cold, and from being very acidic and very basic. Eubacteria live in less harsh conditions than that of Archaebacteria. Bacteria are found in everything around us and can be helpful to some organisms, and also very dangerous to other organisms.

Protists are either uni or multicellular organisms of the Protista kingdom that live in wet, or moist conditions. Protists, unlike bacteria, are eukaryotic. This means that the organism has a membrane-bound nucleus and other membrane-bound structures in their cytoplasm. Protists are both autotropic (synthesize energy on its own) and heterotropic (obtaining energy from organic compounds created by autotrops) depending on its specified species. Protists come in many different forms including Ciliates, Sarcodians, Zooflagellates and Sporozoans. The article focuses on different classifications of Ciliates and the main structures that compare and contrast each classification.

Ciliates

Ciliates in the phylum Ciliophora are protozoans that are covered in short hairlike projections that are called cilia.

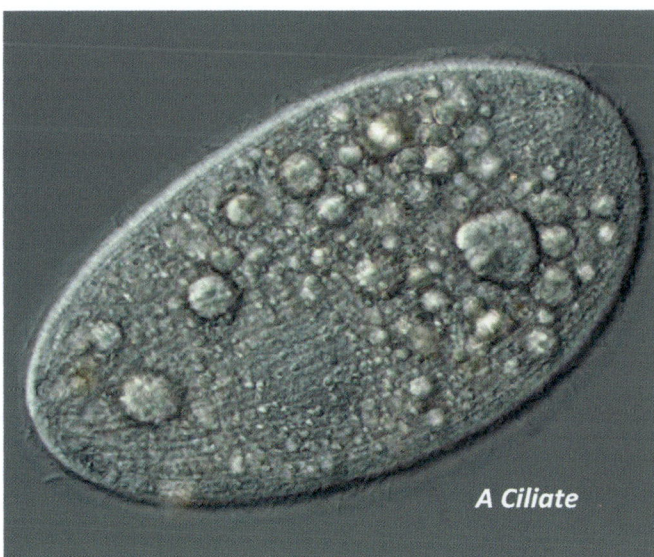

A Ciliate

Protozoa have cilia to move through water, feed, attach onto things, and feel. A distinguishing factor of Ciliates is the presence of two nuclei; the macronucleus and the micronucleus. The macronucleus contains short pieces of DNA and it also pinches off during cell division. During cell division, the cell will completely separate into two identical organisms, thus explaining the term "pinching off". The micronucleus contains two copies of each chromosome. There may be several micronuclei throughout the organism. During cell division, the nucleus will divide through mitosis.

Ciliates have contractile vacuoles and sometimes digestive vacuoles. The contractile vacuole is used to push water out of the cell. If the contractile function of a cell is ruined, the cell will swell until it bursts. The digestive vacuole is the part of the cell that digests food and utilizes nutrients. The digestive vacuole is connected to the mouth, or the cytosome. The cytosome is sometimes set back in a grove in the cell called the oral grove. Cilia that are placed around the oral grove generate water currents that help move food into the cell's mouth. These organelles are an essential part of the cell's structure and make sure the cell's stay healthy and alive.

Location/Collection

The specimens documented were retrieved from the Genesee River next to Rochester Institute of Technology. The source of the Genesee River starts in Potter County, Pennsylvania, and flows 160 miles north, where it ends in Lake Ontario. The river flows through different waterfalls and gorges in Letchworth State Park, and through the City of Rochester. There are three waterfalls the river flows over while in the City of Rochester: Highfalls, Middle Falls and Lower Falls. There are many different species of micro-life living in the Genesee River, which can easily be documented if samples are collected properly.

When collecting samples of water from the river, it is important to bring a part of the habitat back with the sample, including an aquarium with riverwater and plant life. Having the plant life will make it much easier to acquire live and active specimens by allowing the organisms to feed off of and attach to the plant life in their habitat. Mosses, grasses and soil samples are a good place to start to see living and moving specimens.

Handling/Preparation

After retrieving the sample, it needs to be prepared a certain way to make viewing easier. When extracting a water drop, make sure to take the sample from around/inside of the moss or plant life in the aquarium. Taking the sample from the plant life usually guarantees more organisms that come along in the water drop. Cells typically stay around plant life and algae because that is their source of food. Adding a piece of plant to the slide will make it easier to find more life. Some micro life likes to attach to a plant leaf and filter feed in the water. The more simulation of the cell's natural habitat, the easier it will be to view the subjects.

When it comes to photographing, it is tricky to get the subject to stay in place. One way to make the subjects stop moving is to add glycerin to the water sample on the slide. Glycerin will slow down or kill the cells, and retain their structure to a degree. Using other

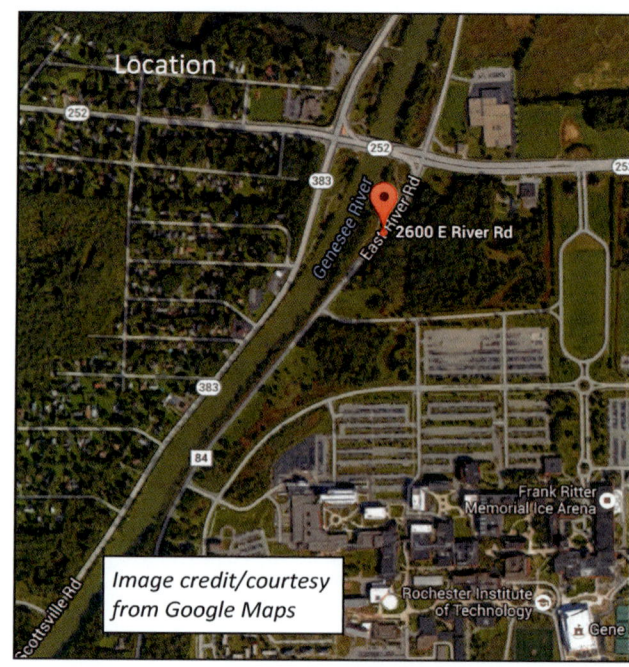

Location

2600 E River Rd

Preparation of specimens

mediums to kill the subject, like any alcohols, will make the cells shrivel up and will not be up to standards for photographing. The easiest way to add glycerin to the sample is to add it directly to the edge of the cover slip on the slide. The glycerin will flow underneath the cover slip and will diffuse through the water, creating a glycerin solution.

Using too much glycerin can end up shrinking the cells and destroying its structures. Too much glycerin can also push the subject to the edge of the slide, creating an unphotogenic mess. A good way to visualize using too much glycerin is to envision an overflowing river gushing through a town. The river will take any debris in its path and create a cluttered mess in the path of the water. There will be nothing left to view on the slide because everything will be moved to one corner. To avoid this, make sure to use a small dropper with minimal glycerin,

and be aware of where debris is in the slide. Doing this will prevent debris from flowing into the subject and into the photograph.

Euplotes Eurystomus
Class: Spirotrichea
Subclass: Hypotrichida

Important/Unique structures:
(See following pictures).
[1] Cirri: used for walking/swimming
[2] Macronuleus region
[3] Micronucleus region
[4] Contractile Vacuole
[5] Cytosomal Region
[6] Oral Grove

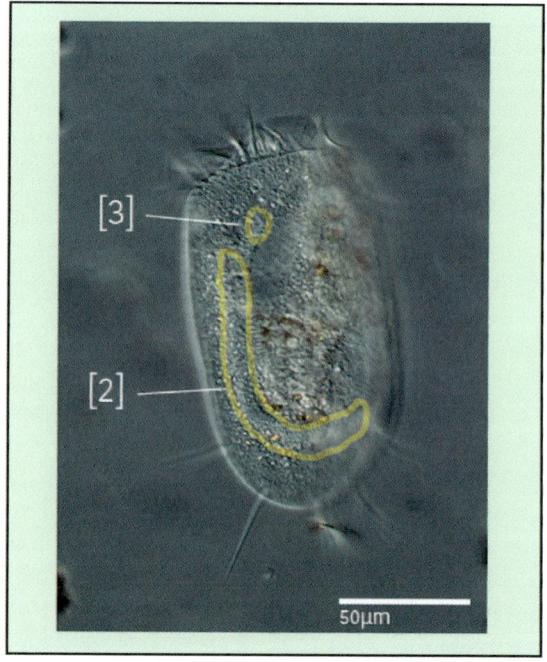

[3]

[2]

50µm

Additional Information

Similar to cilia functions, the cirri are used for walking and swimming.

Euplotes eurystomus is a carnivorous ciliate and an a commonly used food source in research studies is the Tetrahymena organism.

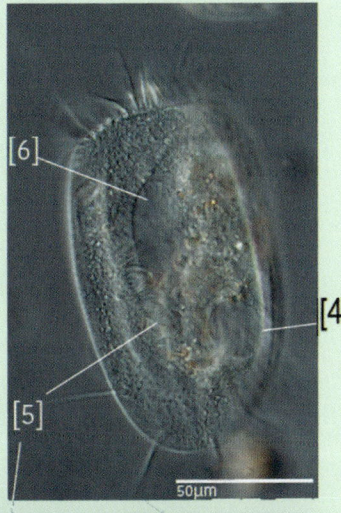

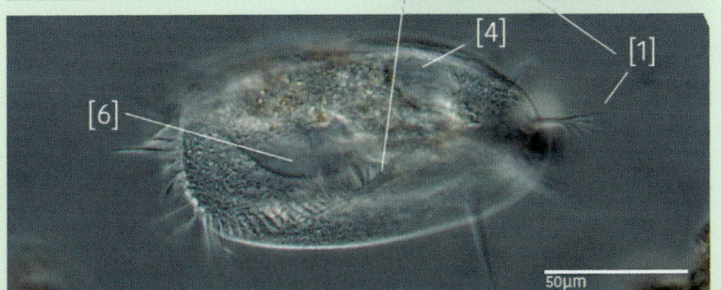

The Author

Jason Dinelli, as of December 2015, is a third year Biomedical Photography student at Rochester Institute of Technology expecting to graduate in May 2017. His concentration is in high magnification imagery and is striving to get a job in the microscopy field. Jason always had a passion for photography and biology and is now combining both of them through the Photographic and Imaging Technologies major at RIT.

Contact Information:
jdinelli125@gmail.com

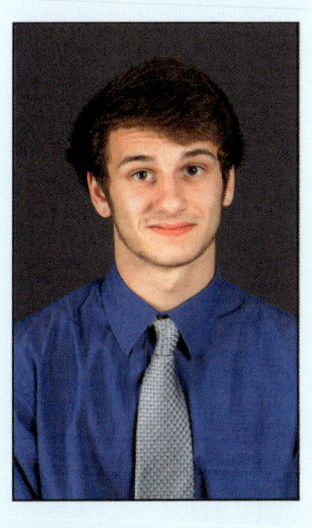

Holophrya discolor

Class: Prostomatea
Subclass: Prostomatida

Important/Unique structures:

[1] **Mouth**
[2] **Cytopharyngeal Basket**
[3] **Macronucleus**
[4] **Micronucleus**
[5] **Contractile Vacuole**
[6] **Fat Droplets (yellow)**

Additional information:

• The cytopharyngeal basket is veinated in structure and lined with cilia to help move food through the organism.

• Holophrya discolor may have more than one micronucleus, and it also may have no micronucleus at all.

• This organism is an algae eating ciliate.

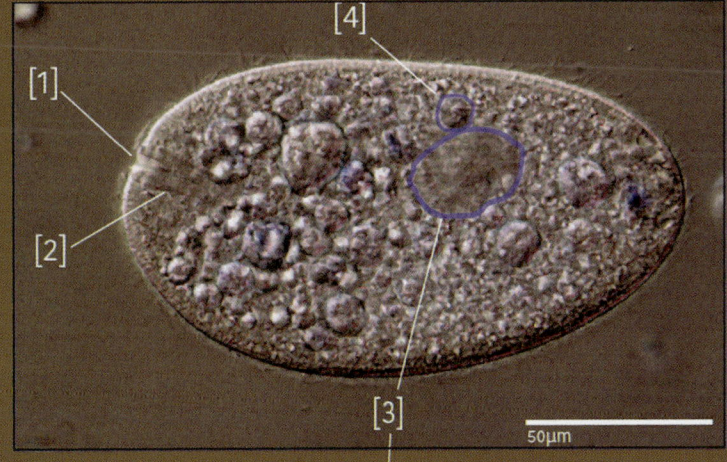

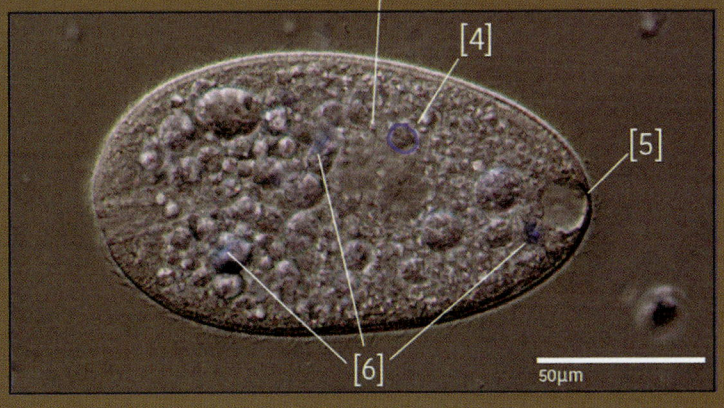

Urocentrum turbo

Class: Oligohymenophorea
Subclass: Hymenostomatida

Important/Unique structures:

[1] **Mouth**
[2] **Medial Belt**
[3] **Macronucleus**
[4] **Micronucleus**
[5] **Contractile Vacuole**
[6] **Tail Cirrus**

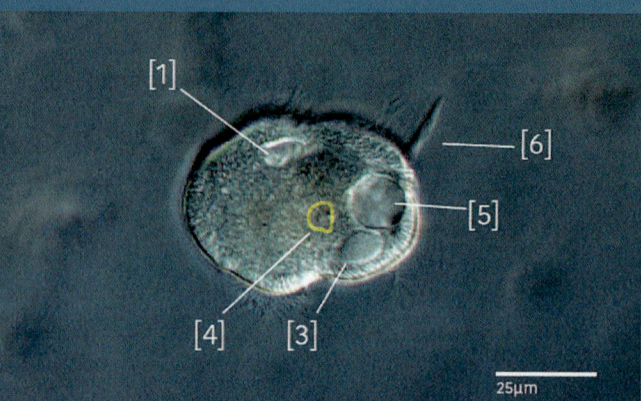

Additional information:

• The medial belt is used for increased propulsion through water and it explains why the organism is usually spinning when observing it.

• The tail cirrus is used for swimming and directing the organism.

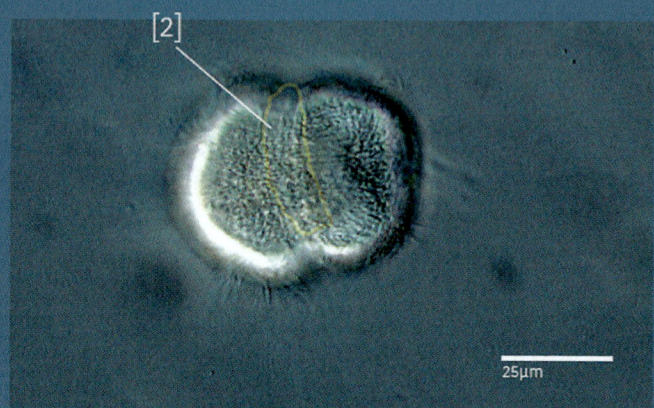

Strombidium caudatum

Class: Spirotrichea
Subclass: Oligotrichia

Important/Unique structures:

[1] **Peristome**
[2] **Cirri**
[3] **Macronucleus**
[4] **Food Vacuole**
[5] **Contractile Vacuole**

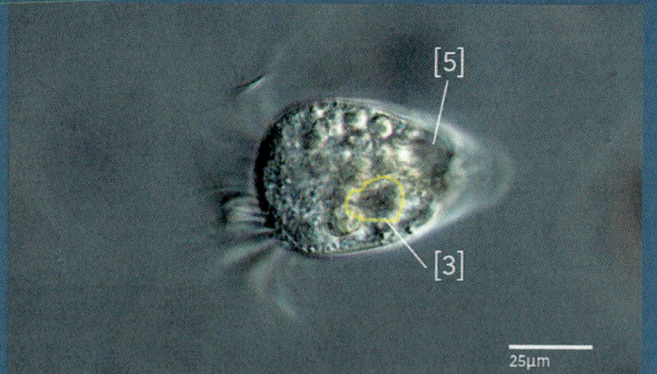

Additional information:

• The peristome is the protruding oral region of Strombidium caudatum. The mouth is at the posterior end of the peristome

• There is an entire ring of cirri around the front of Strombidium caudatum. It creates currents to move food into the mouth of the cell

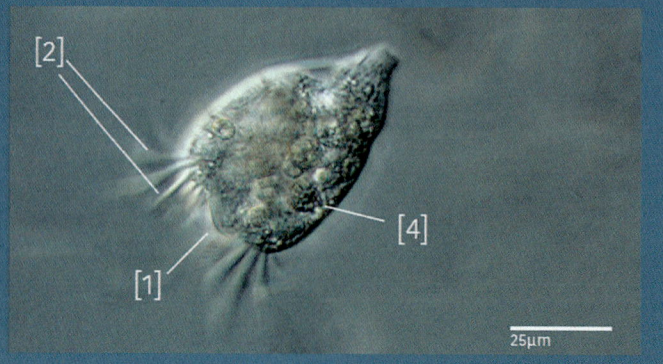

References

http://www.digilibraries.com/html_ebooks/109901/18320/

www.digilibraries.com@18320@18320-h@18320-h-1.htm

http://plankt.oxfordjournals.org/content/33/7/998.full

https://en.wikipedia.org/wiki/Ciliate

http://www.funsci.com/fun3_en/guide/guide1/micro1_en.htm

http://www.uic.edu/classes/bios/bios104/mike/bacteria01.htm

http://pinkava.asu.edu/starcentral/microscope/portal.php?pagetitle=assetfactsheet&imageid=26938

http://jcs.biologists.org/content/joces/3/4/493.full.pdf

http://jcs.biologists.org/content/joces/15/2/379.full.pdf

http://pinkava.asu.edu/starcentral/microscope/portal.php?pagetitle=assetfactsheet&imageid=26898

http://c2.griffithps.schoolwires.net/cms/lib07/in01000714/centricity/domain/49/chap12.pdf

http://www.cityofrochester.gov/geneseeriver/
http://eol.org/info/456

http://www.ucmp.berkeley.edu/protista/ciliata/ciliatamm.html

http://pinkava.asu.edu/starcentral/microscope/PDF/buildPDF.php?imageid=27343

http://www.plingfactory.de/Science/Atlas/KennkartenProtista/01e-protista/e-Ciliata/e-Ciliata1.htm

http://www.microscopy-uk.org.uk/mag/indexmag.html?
http://www.microscopy-uk.org.uk/mag/wimsmall/cilidr.html

http://teachers.henrico.k12.va.us/godwin/strauss_s/sscwebpage/tutorials/protista%20tutorial.pdf

http://jcs.biologists.org/content/joces/1/4/439.full.pdf

Information

All images were taken on the Zeiss Axioskop 2 MOT with the AxioCam HRc camera using differential interference contrast (DIC).

Contact Information:
jdinelli125@gmail.com

Article web address:
www.microscopy-uk.org.uk/mag/artdec15macro/Dinelli_Jason_Final_Article.pdf

Other articles by the students at the **Rochester Institute of Technology** are located online at:
www.microscopy-uk.org.uk/mag/artdec15macro/

Editor's Note:
This article is one of many published in Micscape annually on behalf of students attending a 2 year course **for the 'Principles and Applications of Photomacrography'**, offered in the **Biomedical Photographic Communications (BPC) program 2015** (*http://cias.rit.edu/*) at the **Rochester Institute of Technology** (RIT), NY State, USA. **(***http://www.rit.edu/***)**

Micscape Acknowledgements: Many thanks to all the students over the years for sharing their enthusiasm and skills on such a wonderful variety of topics. Writing an article for Micscape initially formed part of their course as an assignment, which was a neat idea originally proposed by Ted Kinsman the course instructor in 2004 and repeated in most following years. Thanks to Ted for all the work behind the scenes to make this happen. Also thanks to Michael Peres, the dept. head (and course instructor 2009) for his introduction and for maintaining the department's generous collaboration over the years.

BPC Program Overview by Professor Michael Peres, *(www.microscopy-uk.org.ukmag/artnov09macro/) BPC_overview.htm)* department head.
Visit Professor Peres' website
(*https://people.rit.edu/mrppph/*)
Contact the Course tutor 2015 Ted Kinsman
E: emkpph@rit.edu

'From the review of the BPC programme by Prof. Michael Peres published in Micscape, Nov. 2009.'

One of the glorious aspects of a microscope is that it can open up to us hidden dimensions in ordinary items around us and even give us a richer understanding of the complexity of reality.

Microscopists, very often and quite legitimately, want to look inside things, take them apart, cut sections and reveal secret wonders. In this essay, however, I want to look at the outsides of some things, some surfaces—up close. I'll show you an image or 2 or 3 of an object and sometimes let you meditate on it before I tell you what it is. I've put the answers further on in this article so as to not spoil it for you!

 **1**

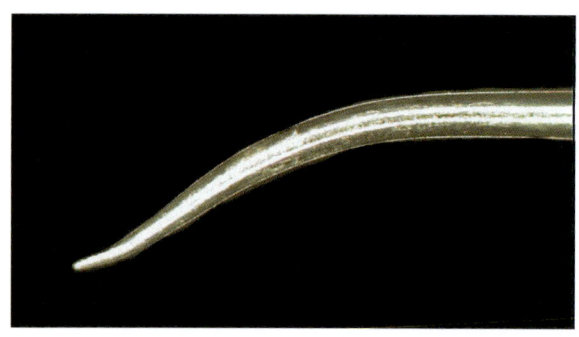

Another... **2**

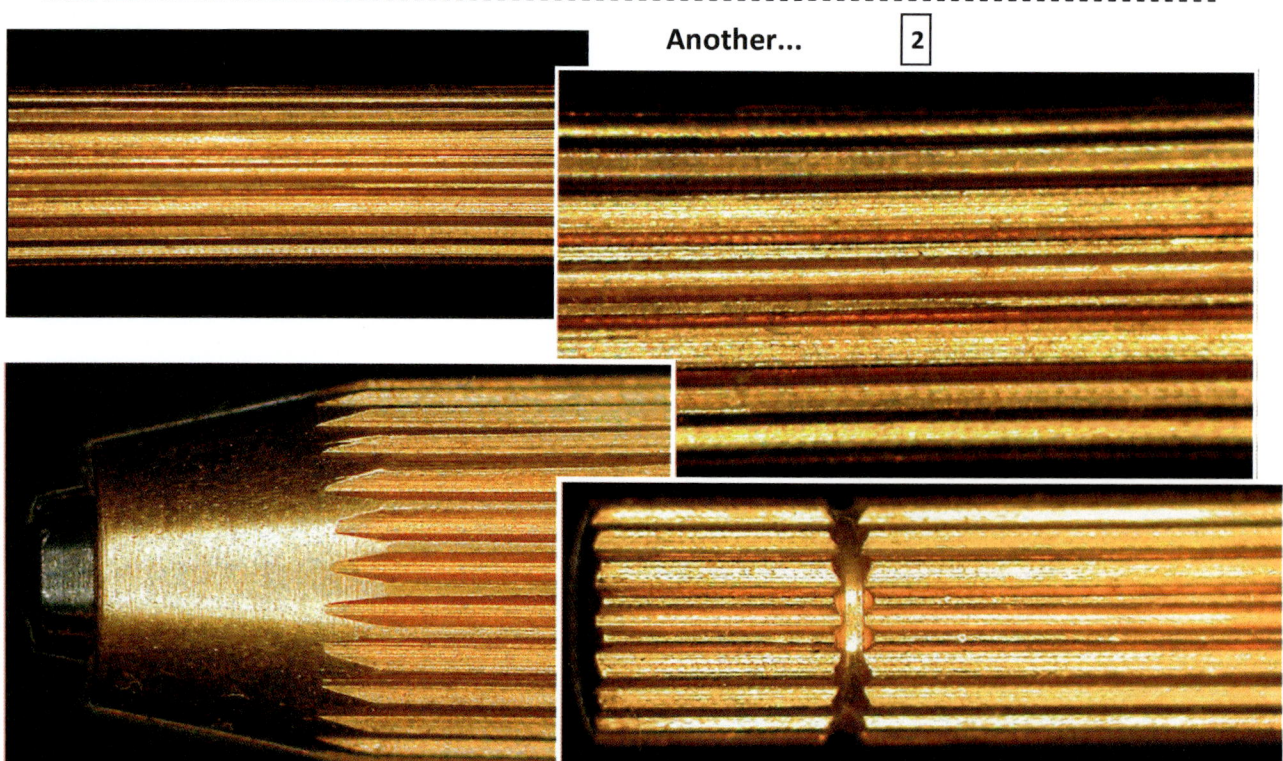

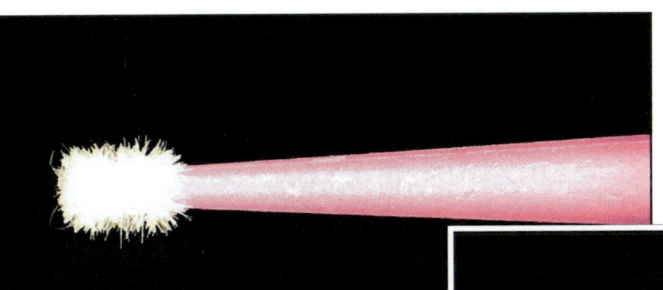

You can find these on Amazon among other places. | 3 |

No. Not a toothbrush.

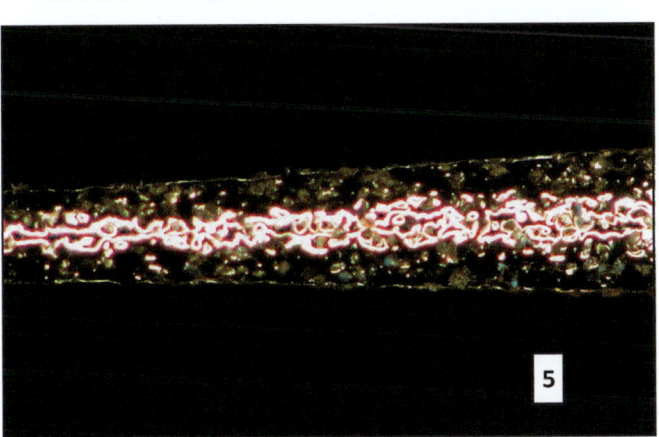

| 4 |

| 5 |

| 6 |

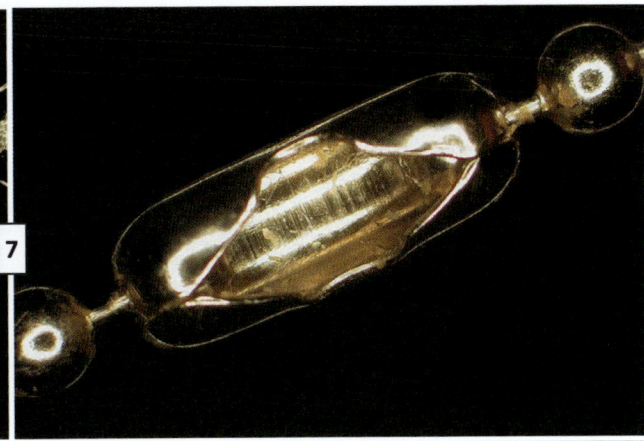

| 7 |

This is a dental probe; the first 2 images are of the handle and the third one is of the tip. I often use a pair of them as dissecting needles. They are very sharp as my left index finger will attest (and, NO, I wasn't attempting vivisection–I hate pain)! Yet, when we look at that tip magnified, it doesn't appear to be

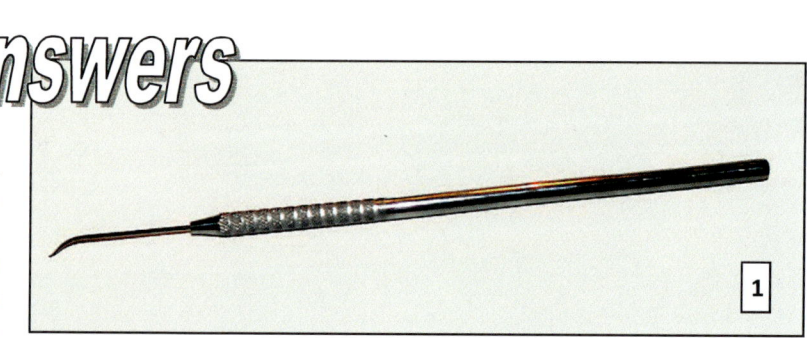

all that sharp–a lesson in perspective. The texturing on the handle is not mere decoration. A smooth steel handle, especially when wet, can be very slippery. The texturing helps one keep a firm grasp on the probe.

The first 2 images are of the handle of a pin vice, the third shows the chuck or collet, and the fourth is the end of the handle. I have several different sizes and styles of pin vises. I use them to hold different sizes of insect pins and also Minuten-Nädeln which are those tiny, thin pins used for

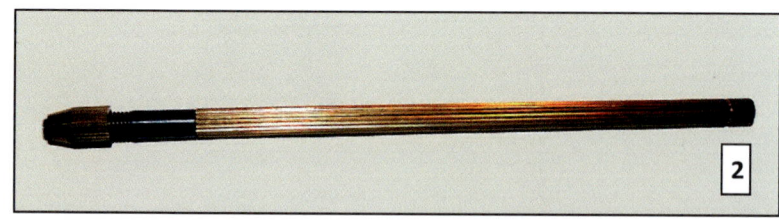

gnats, mosquitoes, etc. You should be extra careful in handling insect pins; they are hellishly sharp and the Minuten-Nädeln, in particular, which because of their small size, could get imbedded in a finger and cause real problems. I always use forceps to load them into a pin vise. What you now have are micro-dissecting needles made of very springy steel. These are exceptionally useful for tasks such as sorting diatoms, forams, or for separating and isolating delicate structures in or on small organisms.

I recommend buying them in quantities of at least 100, since if you buy them in small quantities from craft sites, they are much more expensive. They are advertised both as cosmetic brushes (for putting mascara on the eyelashes of your bumblebees) and as dental swabs (for cleaning the teeth of your

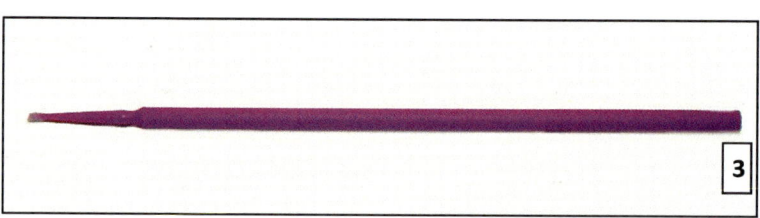

dental swabs (for cleaning the teeth of your sea urchins). Actually, they are incredibly handy and have a variety of uses, one of the most important of which, for me, is cleaning grimy 19th and 20th Century slides. I moisten a micro-brush with the solution that I use for cleaning oculars and objectives and then wipe gently, with circular motions starting at the centre of the cover glass. If there are Balsam flecks or ones from another kind of mountant, then you need to remove these first *delicately* using a micro-scalpel (which we'll consider next). When using the lens cleaner on the brush, be sure not to have an excess that might leak under the cover glass and damage the specimen.

Really good micro-tools, such as those made by Circon, are expensive and should be purchased only if you are going to be doing exceptionally delicate work and are comfortable handling such tools, because they are quite fragile and can easily be damaged if mishandled. There are some inexpensive micro-scalpels

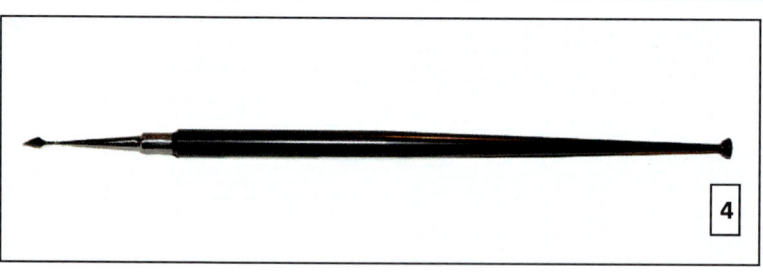

which can be repeatedly sharpened, but it is difficult to get a good edge on them. For many purposes, standard disposable scalpels with a #11 or #12 blade are satisfactory. If you are a bit of a daredevil and take proper precautions– *especially to protect your eyes* –you can make your own, if you have the proper tools. You can break double-edged razor blades into small pieces using pliers or cutters and select those which can be mounted firmly into a small dowel or, if you get lucky, into a pin vise. Another possibility is to take dissecting needles, dental probes, or spear-point dissecting needles and using an inexpensive diamond hone, shape and sharpen to fit your needs.

Useful tools are small diamond routers (and, no, they don't make good presents for your wife, mistress, or girlfriend). They are handy, however, for smoothing edges of calcareous specimens which you wish to make macro-mounts of.

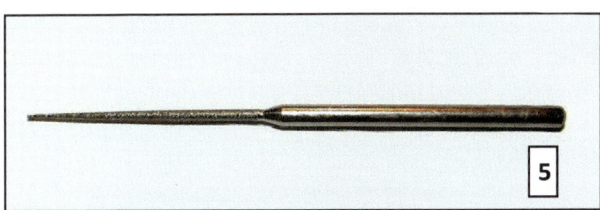

As you can tell from the images, the abrasive is not particularly fine. As a follow-up, you can use an emery board. In this close-up, you can see what almost look like flattened sand grains.

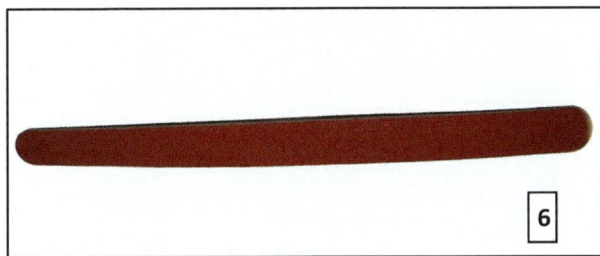

I wear glasses (bifocals) and I have never been very comfortable wearing them while using a microscope, but then that means that I can't read slide or jar labels and would have to constantly be putting my glasses on and taking them off when working in my lab, were it not for my magnifier on a neck chain. It has a 4x lens and makes things much simpler. The chain is quite interesting under the stereo microscope with its reflections from the ring illuminator. The whole assembly looks like this. (Left).

Editor's note: this article which is part 1 of 2 has been abridged here. The full articles can be viewed in full online at: *www.microscopy-uk.org.uk/mag/artapr16/ rh-surfaces-1.html*

Richard can be contacted at:
E: **tunicate@wyoming.com**
His personal web site, full of interesting content is at:
http://rhowey.googlepages.com/home

Surfaces: Part 2 - Specimens.
by Richard L. Howey, Wyoming, USA
Published April 2016

As some of you know, I am obsessed with invertebrate skeletal structures and whenever I get a new specimen, the first thing I do is examine its surface both for its structure and to see if there are any interesting critters which have attached themselves. This time, I want to look at the surfaces of some specimens and also the surfaces of some of their substructures. After all, all we every really see is surfaces according to some analytic philosophers, such as, H.H. Price (who was something of a minor loon). If we slice a tomato in half we are still just seeing a surface. But then as Cicero said: "There is no opinion so absurd but that some philosophy will express it." However, for my purposes here, we shall look at the surfaces of some things that are inside organisms or slices from structures inside organisms.

Let's begin by looking at the shell of a heart cockle. The upper surface is delicately coloured in pastels and has a lovely shape.

The under surface is also intriguing for it has a stark white background with greenish-gray splotches scattered around it. Cockles are a popular food in certain coastal areas. They are boiled and served with vinegar and pepper.

However, if we turn it over and look at the underside, we see the five grooves for the tube feet and indeed we do have pentagonal symmetry.

This next item is not edible, at least, not in its present form. It is a strip of native copper which has, as you can see, oxidized in vivid ways.

Now, if we go back up to the image of the upper surface and look carefully, we can see 5 points where those grooves fold up over onto the top thus hinting at what we will find below.

While we're out wading around in tide pools, we remember that a bit of caution is called for because there are some very spiny relatives of starfish which, if stepped upon can cause a nasty puncture wound. These are, of course, sea urchins, although not all species have sharp spines and some have very eccentric spines indeed. I will show you an example from a tropical urchin *Goniocidaris mikado*. First, 5 spines, which lets you see them from several different angles. Definitely not your ordinary urchin spine! The top looks like a rather fanciful, if not very utilitarian umbrella and also intriguing is the ring around the base just above the round joint which fits into the socket thus forming the ball and joint structure that anchors the spine.

Next, back to the ocean. Starfish have 5 arms–right? Well, not all starfish. I have specimens with 4 arms and 18 arms and 24 arms. There are also some that might appear to have no arms at all. These are the cushion stars and below is an image of the upper surface of one.

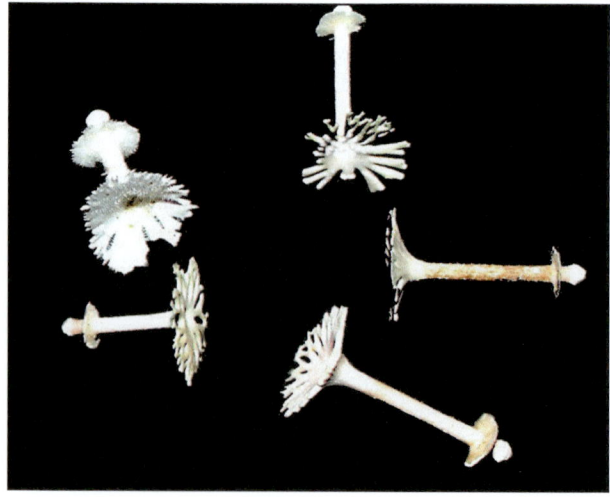

A top view shows us just what an unusual structure this is.

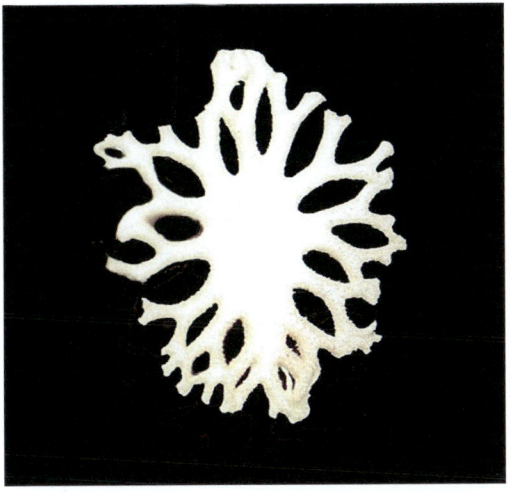

And, a view from the bottom reinforces that notion.

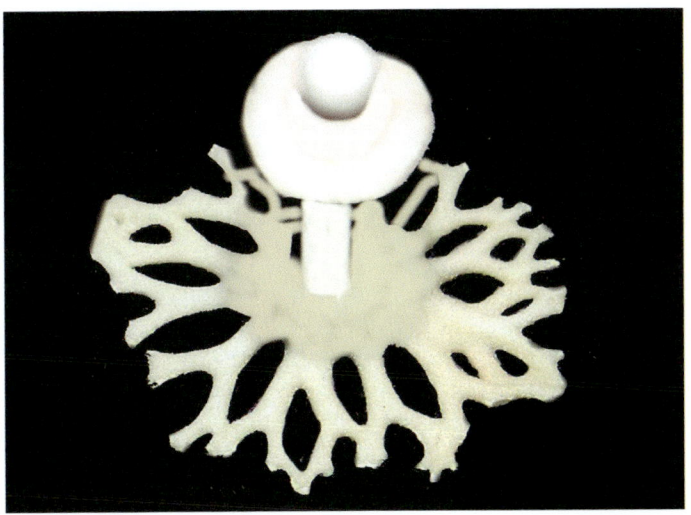

If you want some additional examples of weird sea urchins spines, you can find some real gems here: (*www.microscopy-uk.org.uk/mag/artsep15/ rh-echinoderm.html*)

Almost anyone who collects shells is tempted by the nautilus which is extraordinary not only in its outer appearance, but also because of its intricate internal symmetry which is very beautiful. Here is a view of the shell intact.

It has an almost otherworldly appearance and the organism (*http://www.montereybayaquarium.org/ animal-guide/octopuses-and-kin/chambered-nautilus*) which lives in this marvellous construction most certainly looks like an alien being.

You will notice that on that page, the title is "Chambered nautilus" which you might find somewhat puzzling if you have never seen the sections that have

been cut revealing the internal structure which is simply amazing. Here is an image of one which has been sliced.

What is most striking is the beautiful spiral formed by the chambers. This is known as the Fibonacci spiral and is mathematically described as a logarithmic mapping of the numbers 1, 1, 2, 3, 5, 8, 13, 21, 34 etc. where beginning with 1, 1, all numbers afterwards are formed by the addition of the two previous numbers.

This pattern has been argued to occur not only in the nautilus shell, but in everything from sunflowers to spiral nebulae. As the animal grows it adds new chambers by forming a septum which seals off the small chambers.

One specimen which I acquired was sliced into 3 pieces and the centre section shows the spiral very dramatically and also the siphuncles or ducts that link the chambers. (*See next page*).

siphuncles or ducts
that link the chambers

I also tried a little computer magic on this centre section using the INVERT function in my graphics program and here is the result.

Next, we go from the Sublime to the Beauty Impaired. This unlovely creature is a dried, wrinkled, freeze-

dried nudibranch and it's quite unfair to observe it in this pathetic condition. It belongs to the family *Chromodoridae* and here[1] is a link showing you how vibrant and splendid they can be when living. In fact, nudibranchs are some of the most beautiful of all creatures on this planet.

[1]Go to Google images and search for: *chromodorididae*

So, you might well ask—why acquire such wretched remains as this shrivelled thing? Well, morphologically it is still of interest even in this condition. Nudibranchs often munch on hydroids and manage to do this without triggering the nematocysts or stinging cells. However, they don't digest them, they move them up through their own bodies into the cerata or dorsal projections and they then use them for their own defence. And these are of sufficient interest in themselves to isolate and examine them.

Another freeze-dried marine denizen is a small octopus and from this view we can readily understand why they are called cephalopods (head-foot).

Octopi are amazing contortionists and can move through spaces that seem impossible for them to navigate. The suckers on their feet (arms) are intriguing and are used for food capture among other things. Speaking of food, octopi have a sharp two-part chitinous beak which they use like scissors to rend food and fight off prey.

The suckers are very powerful and can hold prey firmly while the octopus attacks. Here is a close-up of some of the suckers on the arms.

Hidden inside many shells is another dimension of beauty which can only be discovered by slicing the shells into sections. These days such shell slices are produced in quantity and in wide variety. They are often found in craft stores and if you are ambitious and creative you can use them, along with lots of other tiny whole shells, to make Sailor's Valentines which have a fascinating history which you can check out here (*https://www.google.co.uk/search?q=sailor%27s+valentine*).

I'll show you a few sliced shells I got recently and made a couple of small arrangements of.

Here is a close-up and it is, as you can see, quite spiky.

If we turn it over, we get another yet another view of the oddity of this little monster.

Years ago, I cut some small shells by hand using a jeweller's saw. It is very slow, tedious, and demanding work, but the results are pleasing. However, now you can get a packet of ready-sliced ones of 25 pieces for $3 or $4. Of course, they now use slick power tools with sharp blades and can do in a matter of seconds what it used to take me hours to do.

On the ocean bottoms, one can find small and medium-sized monsters. One such is the sea bat. This term is one which has been applied to a variety of unrelated creatures, but I'll show you one which certainly fits the description; this is a small creature but quite fearsome looking.

A close-up reveals books of gills that look rather like bellows.

59

"Now", as Monty Python says, "for something completely different." As I was coming upstairs to my lab after lunch, I noticed that a couple of my flowers in a vase were shedding some bits and pieces, so I collected a couple of stamens one from an *Alstroemeria* or Peruvian lily and the other from a daisy. The pollen on the anther of the *Alstroemeria* is a brilliant red and looks quite festive.

The pollen on the anther of the curved stamen of the daisy is a rich yellow and makes me think of turmeric..

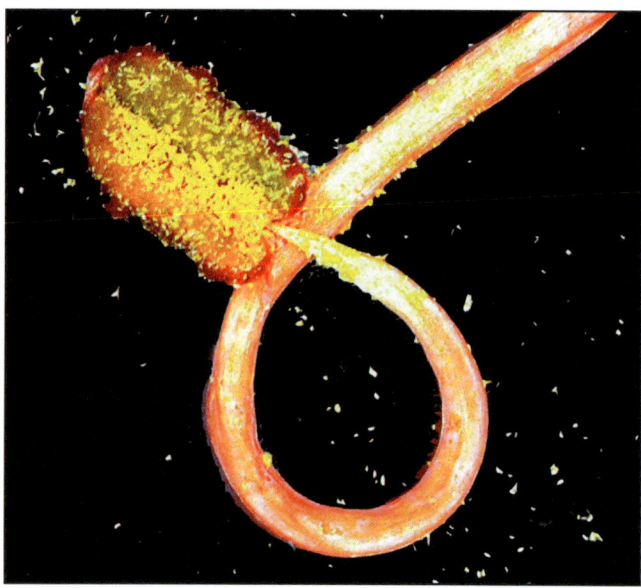

Well, I made it back up into my lab and found a potato. No, it's a cucumber; well, it is a sea cucumber! It is a member of the family *Stichopodidae* and has a ring of 20 tentacles. It tends to live on soft sandy substrates.

And, you must admit, it does look a bit like a potato! The "bumps" on the surface have me curious and it looks like each one sits on a pentagonal area of tissue. This makes me want to start dissecting to see if perhaps there are some dermal or subdermal calcareous plates. But, for now, we'll stick with the surface.

What we will look at next is highly curious, complex, and califragilisitic; it looks rather like someone dropped a small porcelain container that had some saffron in it. Amazingly, this structure has been known for over 2,000 years and was first described by Aristotle and so, is known as Aristotle's lantern.

But why a lantern? Well, this was the first specimen of a group that I unwrapped and it fell apart in this interesting fashion. It permits us to see lots of intriguing

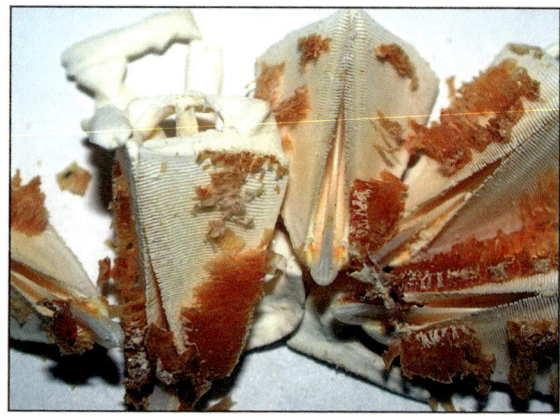

detail including rows of striations and the tips of teeth?

Teeth? Lanterns? Well, look at an intact specimen. First, a top view.

Then, a side view.

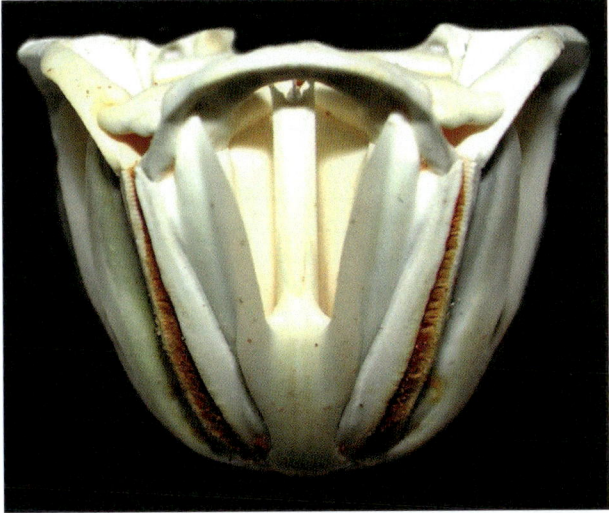

This incredible structure is the feeding mechanism of the sea urchin and usually consists of about 40 parts. Go back up to the broken one and you can clearly see the tips of the teeth. In the other views, you see structures which support muscles attached to the lantern structures and the teeth and move them allowing the urchin to scrape up algae and other detritus for food.

Finally a view of the "test" or shell of a small, lovely, brightly coloured sea urchin, *Coelopleurus maillardi*. When still covered with its spines, you would be unlikely to guess what a magnificent pattern lies hidden below and so the removal of what constitutes one surface (the spines) reveals another surface (the test). All of which should teach us that a multiplicity of perspectives is necessary to even begin to understand the morphology and character of the natural objects which surround us.

Contact the author. E: **tunicate@wyoming.com**
Editor's note: Visit Richard Howey's new website at http://rhowey.googlepages.com/home
This article Online: micscape.prgmag/artapr16/rh-surfaces-2.html

The Origins of Micscape

Micscape began in 1995. Mol Smith (Maurice Smith) had sent a letter to the Postal Microscopical Society[1] which sparked the attention of David Walker. The letter, if recalled correctly mentioned the idea of an Internet site which could bring people together interested in using a microscope as a recreational pursuit. The Internet was still in its infancy in 1995.

Soliciting help from like-minded members of the PMS and the Quekett Microscopical Club[2], the magazine was launched online that November. Many people doubted the significance of the Internet in those early years, and few homes actually had a computer, let alone the 'Internet'.

The style and format of the Ezine was often a topic of hot debate between the early contributors. With Mol wanting a less formal more populist approach, and other contributors desiring a slightly more formal and serious style—democracy reigned, and David Walker was nominated as the Editor and Mol Smith moved back in the electronic shadows to facilitate the technologies required to keep the magazine online and in the public gaze.

Over the following years, many other microscopists kindly contributed articles and material freely to share with the world, with a handful of core writers staying for years to contribute valuable articles and images.

Today, the Ezine and associated web site receive in excess of 4 million unique visitors., and is normally in the top 3 sites listed on Google when a person enters the keyword 'Microscopy'. The magazine remains 'not-for-profit' with financial help obtained by voluntary donations and non-interference sponsorship from a UK microscope seller: **Brunel Microscopes**[3].

Mol Smith created one of the first 30 online shops ever to come online in the UK and was listed in Which Magazine, which proved so popular, he was unable to continue its management. Brunel microscopes now manages the shop to offer support and instruments to the amateur as well as the professional microscopist.

References:
1. *http://myweb.tiscali.co.uk/postalmicsoc/*
2. *http://www.quekett.org/*
3. *http://www.brunelmicroscopes.co.uk/*
Ezine— (Micscape Online): *micscape.org*

My Route to Microscopy and Crystal Pictures
Theo Wyatt, UK
Published April 2016

It is, I guess, a reflection on the narrowness of my grammar school education, that I had never looked down a microscope until some date in the early 1950's. Then, in a chance meeting with my younger brother, with a newly acquired degree in zoology, on his way with his own microscope to his first job as entomologist in Trinidad, he showed me cheese mites, house dust mites and tardigrades. I was intrigued but not captivated.

In 1961 I spotted in a local junk shop a pre-war German toy microscope with a magnification of about x50, and bought it as an educational present for my eldest daughter, then aged 8. Naturally I had to demonstrate how it worked. On the mantelpiece of the girls' play room was a large jar which had once held pickled onions but was now home to a population of sticklebacks, brought back from a weekend at Salisbury with my sister's family. I scraped a little of the green algae coating the inside of the jar and put it on a slide under the microscope. There were two wondrous transparent creatures, brightly illuminated, with two mouths where their head should have been, each surrounded by a rotating windmill of cilia. They were rotifers, found in standing water everywhere; and I was completely hooked. I had to get a real microscope for myself.

I bought *Exchange & Mart* each week, and there a few weeks later I spotted "For sale by tender: three microscopes. Lambeth Group Hospital Management Committee". I walked over to Lambeth Hospital in my lunch hour. Two of the instruments on display were just what I thought I wanted – simple black Leitz machines from the 1930s. I thought I had a reasonable chance of getting one of them if I bid £25 for each. The third was a museum piece in shining brass, bristling with massive knurled knobs, some performing functions I could not discern. I was frankly frightened of it, and feared I should never be able to learn how to control it. I nevertheless risked a £10 bid, and a week later it was mine. It was a Watson *Edinburgh* stand, introduced in the 1890s. I subsequently owned more modern and more sophisticated microscopes, but I spent more hours with and got more pleasure from that old brass instrument than any other. It was also the subject of a truly remarkable coincidence.

We had joined the Merton Scientific Society and there had become friendly with Arthur and Hilda Wyatt. Arthur – no relation – some ten years older than us, was a retired hospital technician and a passionate fossil hunter, whose collection was donated by his widow to the Horniman Museum and is on display there. After one Scientific Society meeting Arthur and Hilda came round to our house and were ushered into the music room where the old microscope happened to be out on a coffee table. As soon as he entered the room Arthur cried out "How on earth did you come to have my old microscope?" It was indeed the very instrument he had used many years before on his first job at Lambeth Hospital.

Arthur gave me a lot of useful advice on microscope technique and I still have a box of specimens he had mounted, but it was another contact made through the Scientific Society who converted microscopy from a passive spectator sport of rotifer-watching into an absorbing creative hobby.

Eric Simpson

He was the brother of Sir Keith Simpson, a forensic pathologist who came to national prominence in several spectacular murder trials, the most noteworthy of which led to the conviction, largely on Keith Simpson's evidence, of John Haigh, who dissolved the bodies of his victims in sulphuric acid, in the erroneous belief that the absence of a body would make conviction impossible.

Eric was the nearest thing to a polymath that I ever encountered; there seemed to be no subject on which he could not provide convincing information. He showed me how to make crystal pictures for the microscope, a skill which he had used, when briefly unemployed in the Depression, to make slides for sale to the microscope shops in London. His last employment was in charge of laboratory equipment at Charing Cross Hospital.

He was also an obsessive hoarder, as we discovered when, after the death of his wife, he left Wimbledon to live near his son on the South Coast. His son asked if we would help to clear his garage before the house was put on the market. The garage was at the bottom of the garden, accessible from the road by a rough track. The door from the garden opened on to a winding canyon, just wide enough for human passage, leading to the centre of the garage, between walls of boxes and cases stretching to the roof. It did not take us long to discover that most of it was rubbish. There was a large tin full of used scalpel blades, many of them blood-stained; there was an even larger tin full of irreparably broken artery forceps; there were several broken X-ray tubes. We hired a skip from a waste-disposal firm, had it placed in the track outside the garage, and started filling it. We had however to be careful, because among the rubbish were some treasures. There were four working microscopes. Three of them were brass instruments contemporary with my Watson. I sold these for him to a dealer through *Exchange & Mart* having spent several hours putting them as best I could into good working order by cleaning, lubricating and adjusting the moving parts and disassembling the objectives and eyepieces and cleaning the tiny internal lenses. When the buyer came to collect them I asked him what sort of customer would be buying them and for what purpose. "Oh" he said, "none of these is ever going to be used. They will all go to smart interior decorators who will put them as conversation pieces in illuminated alcoves in expensive penthouses."

The fourth microscope was an apparently unused Zeiss stand in its box. I bought it for myself because it incorporated a rotating stage, a facility which I had not seen on any other stand and which I fancied might be very useful in making crystal pictures, as indeed it proved to be.

More exciting to me than the microscopes were the cardboard boxes containing more than 100 little bottles and tubes, all meticulously labelled, containing samples of chemicals, some familiar like DDT, some so esoteric that I could not find them in the large chemical dictionary I had inherited from my years in the Chemicals Division of the Board of Trade. Eric had said that he liked to keep samples of all the materials that passed through his hands, and here they were waiting for me to experiment with.

Crystal Pictures

For a year or two I spent a lot of time in the absorbing creation of crystal pictures, trying one chemical after another, carefully scanning the resulting slide for any element of pictorial interest. I slowly developed a clearer concept of what I wanted to produce – a picture with visual impact; a picture with a satisfying feeling of composition; a picture which was more than a random assembly of vivid colours evoking wonder solely because it was produced from a colourless substance; a picture, in short, that an owner might think worth framing and hanging on the living room wall. I found that these requirements seemed most often to be met if the slide contained shapes which could be related to shapes in the natural world.

Equipment

It will be as well if I describe here the equipment I used. All the microscopes I acquired came with simple achromat objectives, and I used the lower powers of these for all my early slides; but as my standards rose, the out-of-focus corners became unacceptable, and I invested in a Russian x3 plan-achromat, which I used exclusively for crystals thereafter. The mechanical stage of the Watson, which had so frightened me in Lambeth Hospital, proved indispensable in the systematic examination of a newly created slide; the combination of mechanical with a rotating stage, which I enjoyed with the Zeiss stand, will, I suspect, be no more than a pipe-dream for most would-be crystal picture makers. I had a Pentax SLR camera body with a screw thread, for which I had to buy a connector to attach the camera to the microscope. Between camera and microscope I fitted a bellows attachment, of the type used for macrophotography, which enabled me, easily and accurately to resize the image to fit the frame.

Enhancing The Image

When the crystal is left to its own devices, the image it produces often contains just too much vivid colour; the eye cries out for light and shade. If you can rotate the slide, each block of colour changes through the other colours of the spectrum and through white and black. There may well be, among all these alternatives, one which is a dramatic improvement on what you started with.

If the crystal does not cover the whole slide, you will need to fill the empty areas with some background colour. The simplest method is to take a sheet of cellophane wrapping from (say) a cigarette packet, and insert it between condenser and slide. You may need to experiment with more than one thickness and by rotating the sheet. When you find a formula that works, you may think it worth incorporating it in a card mount that can be easily manipulated when required in future. And note that whatever colour you add to the background will be added to the whole slide.

Maverick Crystals

Some chemicals can take you by surprise. We all know that a naphthalene mothball, shut up in a drawer full of socks, will eventually disappear completely. The molecules on its surface, when in contact with air can sublime, changing from solid to gas without any intervening liquid phase. Some of those in Eric Simpson's little bottles did the same. One was **styryl methyl ketone**. It produced quite interesting and colourful slides, but if you put them away and looked at them a fortnight later, the slide would be empty; but if you left them for just a few days you would find a totally transformed image. The molecules at the edge of the cover slip, being in contact with air, had wandered off into the wider world. Those further from the open air, now having more elbow room, shuffled around in search of fellow molecules to which they could bind, and in doing so formed large blocks of uniform colour. The result suggested to me the title *Recumbent Figure with Guitar*.

P-chlorphenyl glyceryl ether melted easily but crystallised extremely slowly. Having made the slide and waited 15 minutes in vain for a result to appear, I abandoned it as a failure. Taking up the empty slide a few days later to try another chemical, I was astonished to find the picture I have called *Satellite*. A week later the same slide produced the picture I call *Degas' Ballerinas*.

Salicin melted readily, but in its liquid form was so viscous that it refused to flow by capillary attraction between the slide and cover slip, forming a rope of material which pushed the glass surfaces apart. The result was an intriguing picture, but not one that anyone would want to hang on the wall. If there had been a next time I would have tried to find a solvent.

Unanswered Questions

Let me imagine the first one: "How on earth do I get hold of a sample of benzoyl eugenol?" Brutally frank answer; "Don't even try; it is impossible."

Sour grapes answer: "If you succeeded, you would be very unlikely to replicate my result; much depends on the impurities in the particular specimen."

Philosophical answer: "In this area, chance is king. My brief friendship, half a century ago, with Eric Simpson, and the invitation to help clear his garage, were two outrageous pieces of luck; without either I would never have made a single picture. But they would never have happened if I had not paved the way by joining the local Scientific Society. You have already taken the first step on that path by consulting the website of *Microscopy-UK*. Keep in touch with fellow enthusiasts; chat up any doctors or hospital lab technicians you meet socially; ask the pharmacist at your local chemist if he ever disposes of drugs from which he could spare a few granules for you to experiment with.

Reflections

I do think that, given the care and skill that must be devoted to locating an image in a crystal slide and manipulating it into something of aesthetic quality, the best crystal pictures can justifiably claim to be abstract works of art; but I have to accept that, at least for my lifetime, those who make them will continue to be regarded as geeks, unfit to be mentioned in the same breath as Jackson Pollock and Mark Rothko. Nevertheless, I dream that one day a lucky microscopist will spot an image on the slide that is unmistakably a Christmas robin, and will sell it to a publisher of greeting cards, thus taking the first step in a claim to be taken seriously. And then, who knows? One of us may find a poster-size enlargement of one of our pictures on the walls of Tate Modern.

by Theo Wyatt, UK

Please Note:

"The author is 95 and partially sighted with macular degeneration; correspondence by e-mail is laborious and avoided wherever possible. He is however happy to discuss his work and answer questions on the telephone from readers who will make allowance for his hearing loss by keeping the pitch of their voice low and by speaking slowly and distinctly. The telephone number is 020 8540 2708. Ring any day between 10.00 and 22.00."

Article is online at: *www.microscopy-uk.org.uk/mag/artapr16/tw-polar-crystals.html*

(Image content abridged here. Editor).

Crystal Gallery (taken with cross Polarising filter):

Recumbent Figure with Guitar. (styryl methyl ketone)

Benzoyl eugenol *(Fire in the cathedral?)*

DMEM Stilbene

Hippuric acid

Satellite   (P-chlorphenyl glyceryl ether)

Degas' Ballerinas   (P-chlorphenyl glyceryl ether)

Lauric acid *(Asparagus?)*

Phthalic acid

Hippuric acid

More crystal images for this article are online at www.micscape.org/mag/artapr16/tw-polar-crystals.html

Figure 1. Description in the text.

UNUSUAL MICROSCOPES:
THE ELGEET ZOOM PROJECTION MICROSCOPE
(Ingenious, Practical, and Extinct)

By Manuel del Cerro and Dietmar R. Krause
Princeton, NJ, USA

Published November 2015

Introduction

This article is the second of a series describing microscopes that represent dead-ends in the evolution of the instrument. The first article was published in Micscape in May 2014 [1]; it dealt with the Bausch and Lomb microscope called The Wide Field Tube. Here we will discuss the Elgeet Zoom-Microtar, a very different instrument that unfortunately, shared a similar fate with the Wide-Field Tube.

If we were to tell young people that: "Once upon the time there were microscopes designed to operate as attachments to slide projectors," the assertion would likely elicit the question: "What is a slide projector?" The answer that it was a device used to project photographic slides on a screen, may only lead to a second question: "What are photographic slides?"

The world is changing incredibly fast, so for the benefit of those born around the year 2000 or later, lets first say that photographic slides were pieces of, most commonly, colour film, with a size of 24 by 36 mm, mounted on cardboard or plastic frames. Since their size was too small to appreciate the images directly, the slides were viewed by projecting them on a screen, using a projector. This was an apparatus that consisted primarily of a strong light source, a projection lens, and a mechanism that brought slides, one at a time, into the light path. Figure 2 illustrates the projector and the slides for the benefit of the new and coming generations, and to facilitate the discussion that follows.

Figure 2. A Kodak slide projector with its distinctive "Carousel" slide tray. The Kodak trays held, 80 or 140 slides, depending on type.

Projectors and slides were ubiquitous; they could be found in teaching institutions, hospitals, research and industrial laboratories, and very many private homes. It was natural then that somebody would design a projection microscope that could use the slide projector light source and fit in the place of the (removed) projector lens. The Elgeet Company of Rochester, NY marketed such an instrument specifically to be used in conjunction with slide projectors made by the Eastman Kodak Company of Rochester, NY. The location and the times (circa 1970) were propitious. Rochester, NY, was home to a Bausch & Lomb operation that was producing

microscopes by the tens of thousands, and bringing to the area workers with considerable optical know-how. The Eastman Kodak Company, an industrial giant, was producing slide projectors by the hundreds of thousands. All the elements were there. The Elgeet Optical Company originated from this rich background:

"The Elgeet Optical Company was founded by three young men who had been boyhood friends: Mortimer A. London, then [1946] a lens inspector at Kodak, with David L. Goldstein and Peter Terbuska of Ilex." [3]

The firm's name is an acronym of L, G and T.

Now all this is gone. The Bard said it: "There is a tide in the affairs of men." [4]

Description:

Elgeet, Rochester, New York. Microtar-Zoom projection microscope (fig. 1). This description is based on a microscope that is item #334 in the MdC Microscope Collection, presently at the National Museum of Health and Medicine, Silver Spring, Maryland.

The base is an aluminium cylinder 5.2 cm wide and 6.4 cm tall; it is pen at its lower end to receive and transmit the light emitted by a slide projector. The stage is circular, 4.7 cm in diameter, and it protrudes

1.0 cm from the base. An iris diaphragm is attached to the stage. The stage has a metal holder for standard 2.5 by 7.5 cm glass slides. The arm rises 2.2 cm above the stage and it ends in a horizontally placed ring that supports the microscope body. This ring carries the inscription "Made in U.S.A. by Elgeet Optical Co." The body is 5.5 cm tall when the ocular is pushed fully in. Focus is attained by displacing the body in or out of its supporting ring. The ocular carries the inscription "Elgeet Zoom-Microtar f1.5." Zoom is achieved by sliding the ocular in or out.

The microscope has a leatherette-covered box that carries the inscription "Elgeet Zoom-Microtar" on the outside. The inner surface of the cover carries the inscriptions "Elgeet Rochester, New York" and "Quality is our watchword. Precision Engineering our constant goal."

OPTICAL PERFORMANCE:

At a distance of 1 meter the image on the screen is enlarged 41.5X, with the ocular pushed fully out it can be zoomed up to 65X. The 6 images are very sharp but become dimly lit at distances greater than 1 meter. Designed to function as a projection attachment for Kodak 35 mm slide projectors, and highly portable, this instrument performs very well as a low power compound microscope on its own. All that is needed in this situation is to aim the base to a light source; the skylight works particularly well.

The Elgeet Microtar projection model appears to be extremely rare; in many decades of dealing with microscopes and photographic equipment, and searching for offerings of antique equipment (eBay and other sources), we have never seen another example, neither have we seen advertisements for this instrument. A Google search shows a single hit under the revealing title of "What is it?" [5]

Discussion

Elgeet was not alone in thinking that a slide projector would make a good base for a projection microscope. An example is the well-known Pradovit Micro-attachment produced by Ernst Leitz GMBH, at Wetzlar. [6]

There are similarities as well as substantial differences between the Elgeet Zoom and the Pradovit. The basic concept was the same, to take advantage of the strong light source on which slide projector design was based. The differences however, were significant.

The Elgeet Microtar-Zoom was designed for use with the Kodak slide projectors in the Carousel, amateur" version (Figures 2 above, and 3 below), or the Ektagraphic, "professional" version (Figure 4). The magnification range obtained by zooming was a modest 1.5, from 41.5X to 65X. As noted above, the Elgeet was able to function as a standalone low power microscope.

The Pradovit was designed as an attachment to the Prado slide projector; because of its configuration it was not capable of working independently of its projector. The Prado with objectives from 3.5X to 25X provided a magnification range of 7 times. **(Figure 5) [3]**

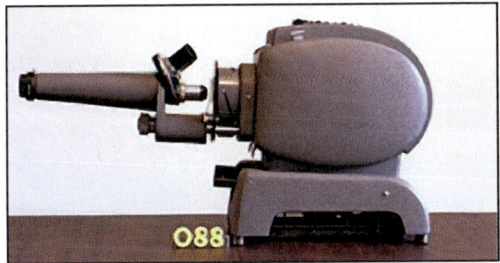

Produced by Ernst Leitz GMBH, Wetzlar, Germany, this instrument is microscope #088 in the MdC Microscope Collection. Originally owned by the Ellis Fischel Cancer Center of the University of Missouri, presently it is part of the Rare Book Collection, Edward Miner Library, University of Rochester, Medical School.

Figure 3. The Elgeet Microtar on a Kodak Carousel projector.

Figure 4. The Elgeet Microtar on a Kodak Ektagraphic projector.

67

The end of the projector-based microscopes was the result of a chain reaction. The demise of the photographic slides as a result of the Digital Revolution brought about the demise of the slide projectors, and this naturally the demise of this type of microscope. This is however, not a story of loss. The projection of slides on a projection screen is now replaced by the display of the photographic image on computer screens (Figure 6). The image can be displayed, enhanced for optimal visibility, stored, and shared across the room or across the world; its features can be selectively enhanced, or automatically quantified for scientific or technical studies. All this is infinitely more than any projection microscope could ever achieve; this is progress.

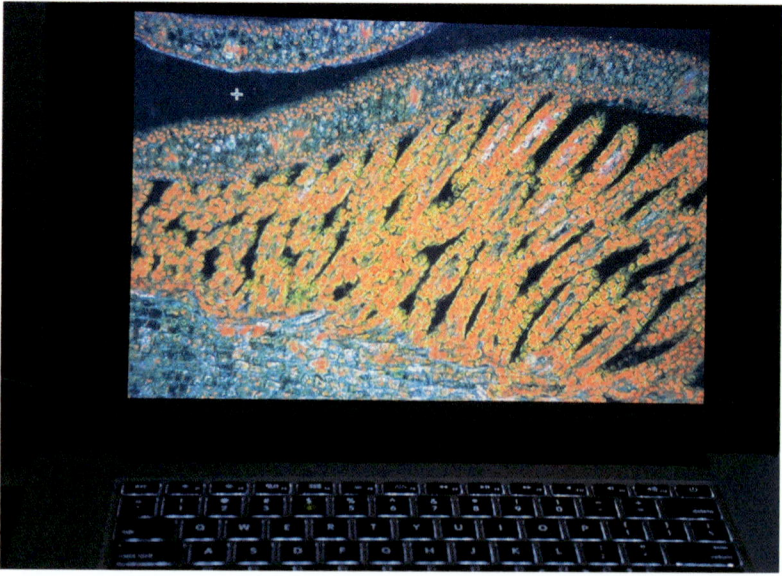

Figure 6. The image of a section of Iris germanica, seen in dark-field and displayed on the Retina screen of a MacBook laptop computer

SOURCES

[1] del Cerro, Manuel and Dietmar R. Krause: Unusual Microscopes: The Bausch & Lomb Wide-Field Tube. MicScape, May 2014.

[2] Kingslake, Rudolf. The Rochester Camera and Lens Companies. Rochester NY, Photographic Historical Society. 1974 <www.nwmangum.com/Kodak/Rochester.html>

[3] 1950-60's PRADO Projector w/ Microscope <Attachment. www.Leitzmuseum.org>

[4] Shakespeare, William: Julius Caesar Act 4, scene 3, 218–224.

[5] What is it?
http://55tools.blogspot.com/2012/09/set-458.html

[6] Projector Images
https://www.google.co.uk/search?q=E.+Leitz,+Prado+Projector

Email author **Manuel del Cerro** at:
poodlesgo@yahoo.com
Published in the November 2015
issue of Micscape Magazine
www.micscape.org
Read this article online at:
www.micscape.org/mag/artnov15/mdc-drk-Elgeet.pdf

Micscape Facts
Micscape enjoyed its 20th anniversary the month the article above was published. The image below headed that month's publication and was created from sources captured by Wim van Egmond of the Netherlands. |Every issue published remains on line. Many articles have been translated into Chinese and are duplicated on a web site in China to encourage the use of a microscope by non-professionals.

Below—Issue no. 20, June 1997., the cover of the first issue since its inception in November 1995 with a discrete index for that month; earlier issues had the same rolling index.

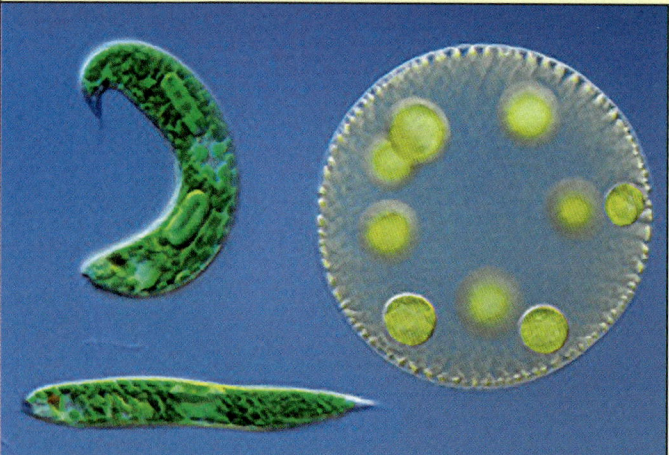

WINTER IS FOR MICROSCOPY III. SOME PROTOZOA

By Anthony Thomas (Canada).
mothman@nbnet.nb.ca

Published January 2016

In Part I, in the April 2015[1] issue of Micscape showing some unicellular algae, I wrote: *"It's now time to take a brief look at some of the multicellular algae, the filamentous algae, and the aquatic protozoa....... to be continued."* Part II (July 2015, Micscape[2]) looked at some multicellular and filamentous algae; Part III considers some of the winter protozoa.

The term "Protozoa" is a taxon of convenience consisting of a large diverse group of animals that are not necessarily closely related.

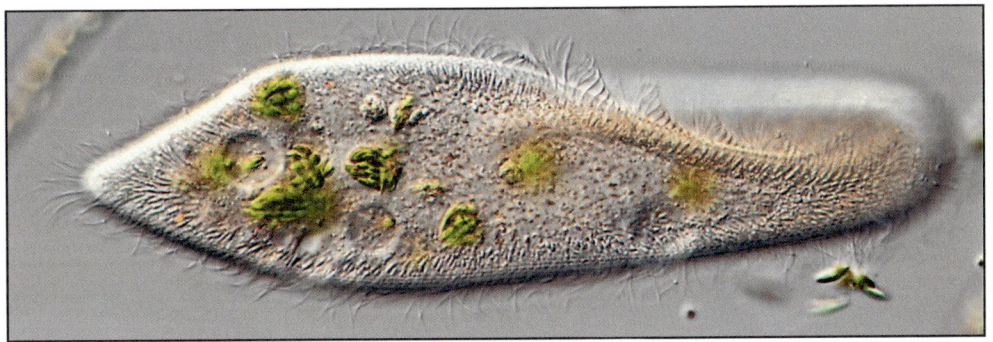

Amoeboids

Included here are the single-celled flexible 'bags' of protoplasm which move by the protoplasm flowing out in the form of pseudopodia; 2 basic types, naked and testate.

1] Naked amoeboids

These species appear to have no fixed external shape, a circular or elongate central area with 'arms' (pseudopodia) sometimes described as a 'stellate morphology' (i.e., star-like).

Figure 1 shows some of their forms and their internal structure. There is a clear outer hyaline cytoplasm (hc); an inner granular cytoplasm (gc); a single nucleus (nu); several food vacuoles (fv), some empty and some with food; and pseudopodia (ps) or 'false-feet'. Three images of live individuals and one image of an individual fixed and stained with a formaldehyde/malachite green mixture. *Fig. 1.*

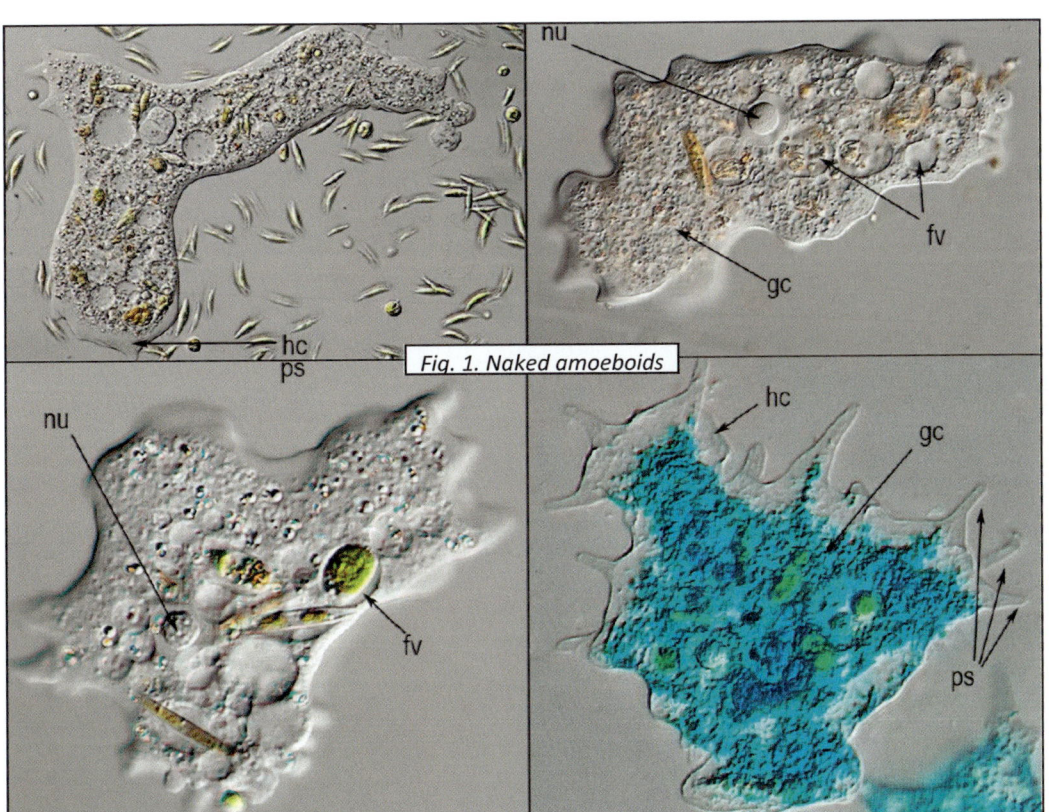

Fig. 1. Naked amoeboids

Most of the naked amoeboid protozoa are usually found 'sliding' over a surface. In contrast, the Sun Animacules (Heliozoa) are free-floating amoeboids. They are unicellular and roughly spherical with many pseudopodia supported by radiating axiopods (*Figs. 2, 3, 4—see next page*).

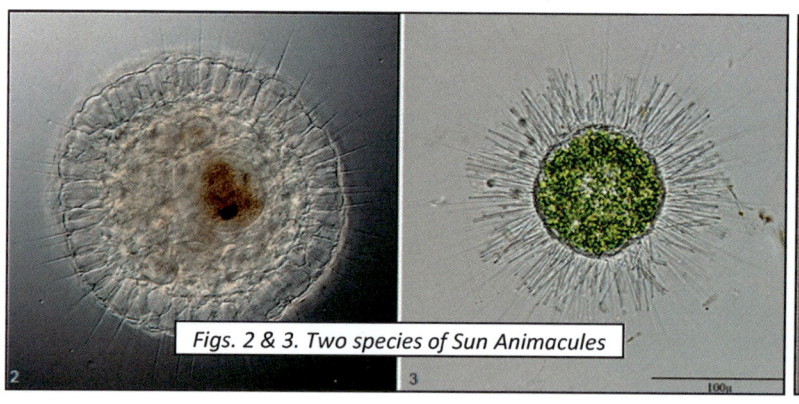

Figs. 2 & 3. Two species of Sun Animacules

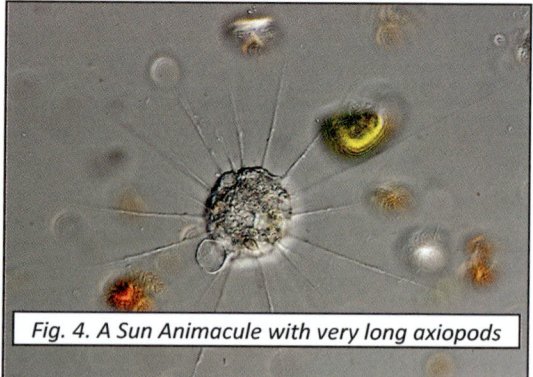

Fig. 4. A Sun Animacule with very long axiopods

2] Testate amoeboids

These little single-cell animals are truly amazing – they construct houses called tests. These tests can be like half a sphere, flat bottomed with a domed top, or vase-shaped. The test may be proteinaceous and is often covered with inorganic particles such as sand or with secreted plates. In either case there is a single opening through which the pseudopodia protrude. The tests are resistant, I often find empty intact tests in samples of pond detritus.

(**Below**)—Amoeboid tests:

a - focused on dorsal surface of an empty test showing the circular bottom opening.

b – same test but focussed on bottom opening.

c – test from the top but focused on the bottom opening showing a live amoeboid within, nu = nuclei, fv = food vacuole, te = test wall

Fig. 5

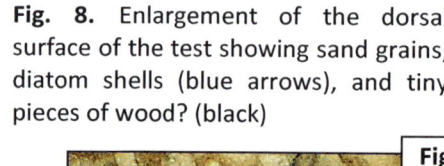

Fig. 6. Living testate amoeboid, focused through dorsal surface of test to show a single pseudopodium (ps) extending from the bottom opening of the test; it is just possible to see the hyaline cytoplasm at the leading tip and the granular cytoplasm behind.

Fig. 7. Test with living amoeboid (am)
a - ventral view looking through circular opening
b – dorsal view showing sand grains, diatom shells and pieces of wood? (black) that constitute the test

Fig. 8. Enlargement of the dorsal surface of the test showing sand grains, diatom shells (blue arrows), and tiny pieces of wood? (black)

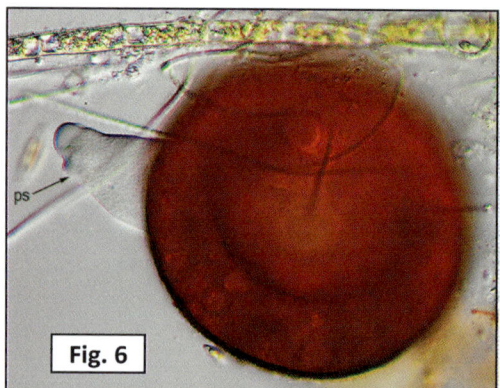

Fig. 6

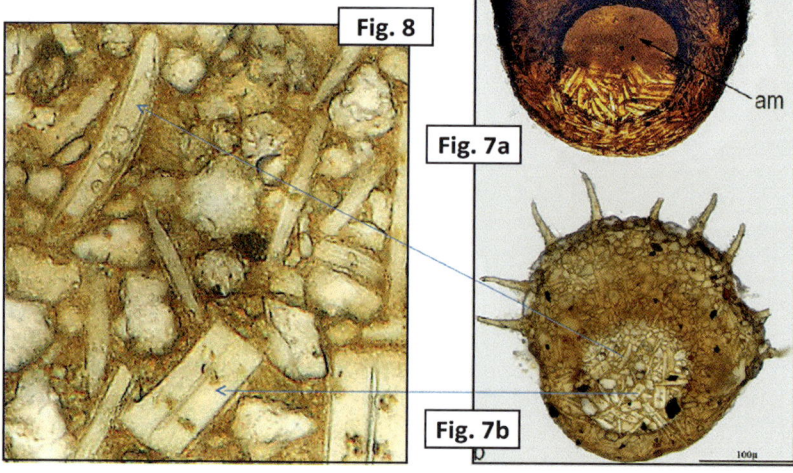

Fig. 8

Fig. 7a

Fig. 7b

Fig. 9a - two testate amoeboids conjugating. Tests appear to be made of sand particles, dorsal view. (*Below*).

Fig. 9b – another pair of conjugating testate amoeboids, focus on slide where pseudopodia are extending.

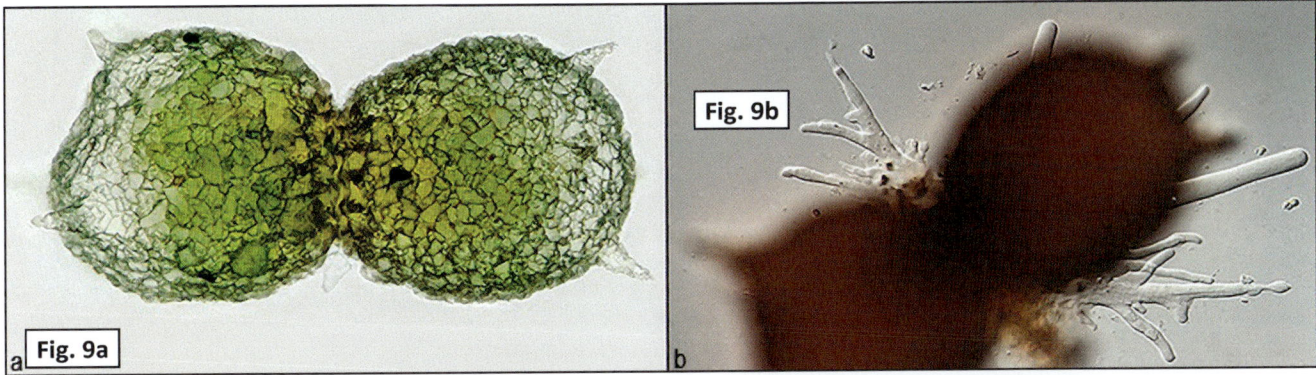

Fig. 9b

a Fig. 9a

b

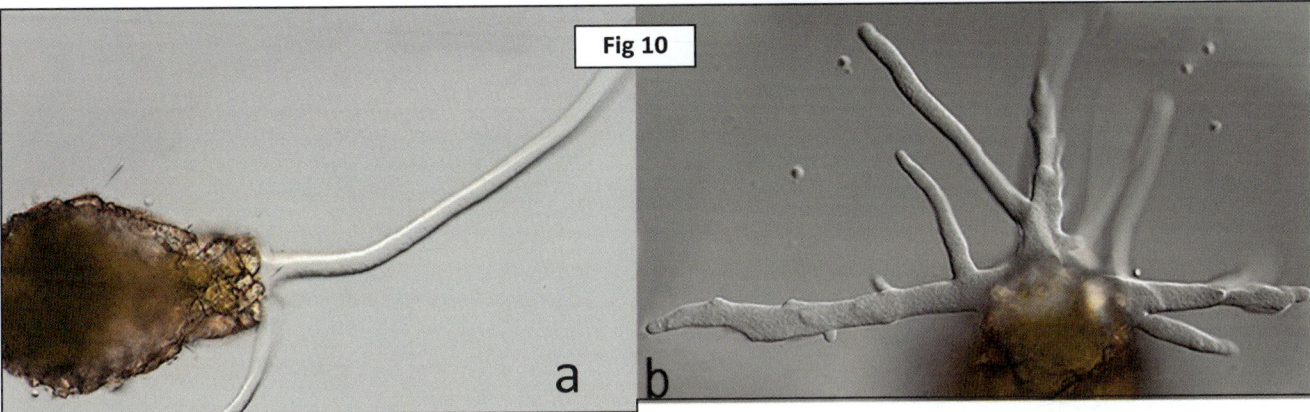

Fig 10

a b

Fig. 10. A vase-shaped testate amoeboid,
a - side view with 2 very long pseudopodia extended
b - same amoeboid on microscope slide with pseudopodia spreading over bottom of cover glass

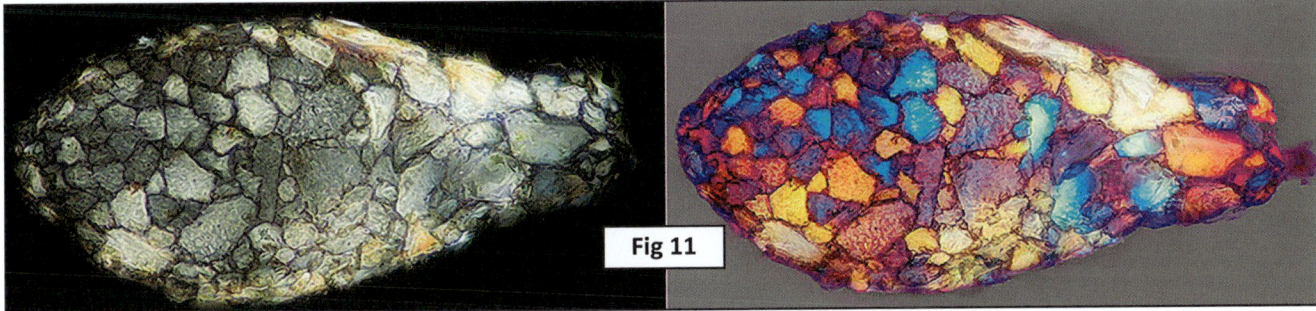

Fig 11

Fig. 11. A vase-shaped test. Left as illuminated with normal lighting, right when illuminated by polarized light. Test appears to be made entirely of sand grains, opening at narrow end top of the vase.

3] The ciliates

These are by far the commonest protozoa in all of my samples. Usually very active and fast swimmers which have to be slowed-down to get a decent photo (Protoslo from Carolina Biological Supply Company works well).

3.1 Paramecia, the "Slipper animalcules", are the 'classic' ciliates being among the first to be seen by microscopists and are widely used in school biology classrooms. Relatively large single-celled protozoa which are just visible, especially when numerous, with the unaided eye. There are several species in the genus *Paramecium* of which *Paramecium bursaria* is readily recognized by its green colour thanks to the endosymbiotic green algae living within its protoplasm

See (Fig. 12).on the next page.

Fig 12. *Paramecium bursaria*, 6 planes of focus

The other paramecia are colourless but are visible under the microscope thanks to their cell wall and the complex organelles within the body.
See this article on line at:

www.microscopy-uk.org.uk/mag/artjan16/at-algae-III.pdf
Comments to the author:
Anthony Thomas. E: mothman@nbnet.nb.ca

A simple differential stain of blood smears using black Quink®

by Chris Thomas, 3 Hall End,
Milton, Cambridge CB24 6AQ.

E: chris@miltoncontact.com
Published September 2015

Published in: Micscape http://www.microscopy-uk.org.uk/mag/artsep15/ct-Quink-blood-stain.pdf

PDF at https://dl.dropboxusercontent.com/u/1646983/Microscopy%20articles/Quink-blood-smears.pdf

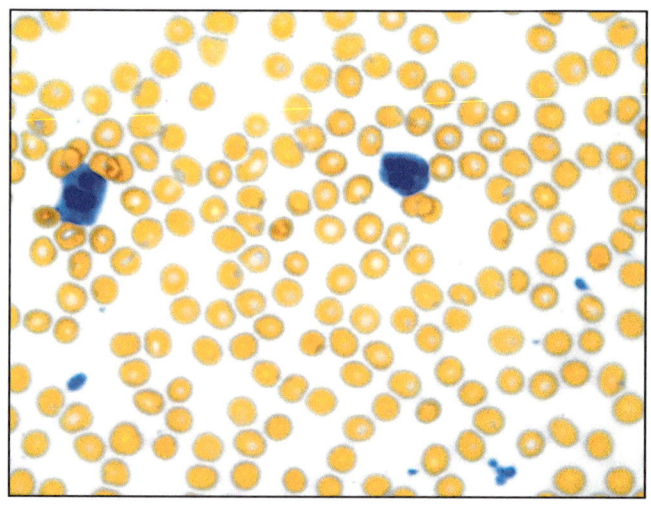

Summary

Dried and fixed peripheral blood smears can easily be differentially stained with Parker's black Quink® fountain pen ink. It successfully increases the contrast of erythrocytes, stained orange/yellow. It differentially stains white blood cells and platelets blue. It is possible to distinguish between neutrophils, small and large lymphocytes, monocytes and eosinophils.

A simple method using common and economical ingredients is given. The method can be used in schools, by amateur microscopists, and by professionals in situations where conventional stains are either temporarily unavailable or unobtainable.

Contents

Summary Introduction
Results
Discussion
Method
Bibliography

Introduction

"What does blood look like under the microscope?" My 82 year old mother asked during a visit, as I was looking at some slides. With a promise to show her, I prepared some smears, yet they were very faint when viewed under transmitted light. Fountain pen ink is a known stain, both of writer's fingers and samples for the microscope. I therefore tried a quick stain with Parker's Quink® black ink and was surprised by the increased contrast obtained, for erythrocytes (red blood cells), platelets and the differential staining of five different types of white blood cells that I could find. Staining was initially variable, so I continued to try a variety of different conditions, based on other researcher's experience and the nature of Quink® ink. The results and discussion of these experiments are given below, followed by the final method. My mother enjoyed her look at the stained blood cells.

Results

An unstained blood smear is clearly seen by eye on the microscope slide, however, the blood cells are barely visible under the microscope (figure 1).

After one minute staining with black Quink®, the walls of the red blood cells (erythrocytes) are sufficiently stained grey or black and the contents appear yellow or orange (figure 2). White blood cells and platelets appear in dramatic blue.

Figure 3 shows examples of five different white blood cell types and platelets viewed with a 100x oil immersion lens.

Figure 1. Left - unstained blood smear on slide. Right – blood as seen at 400x under light microscope.

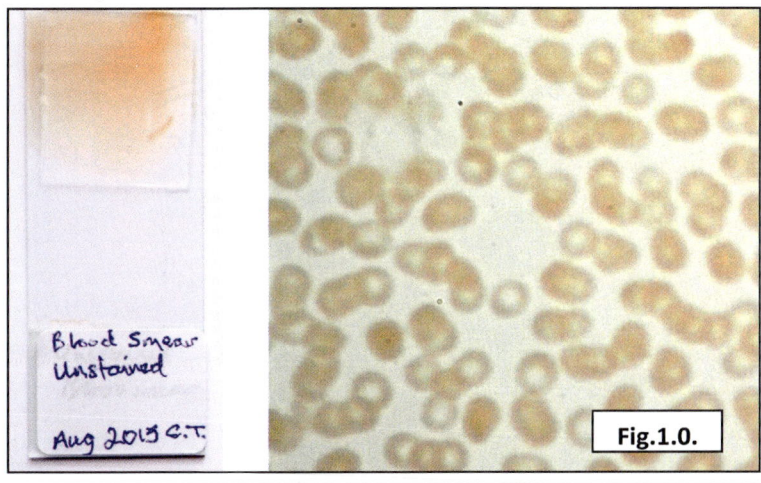

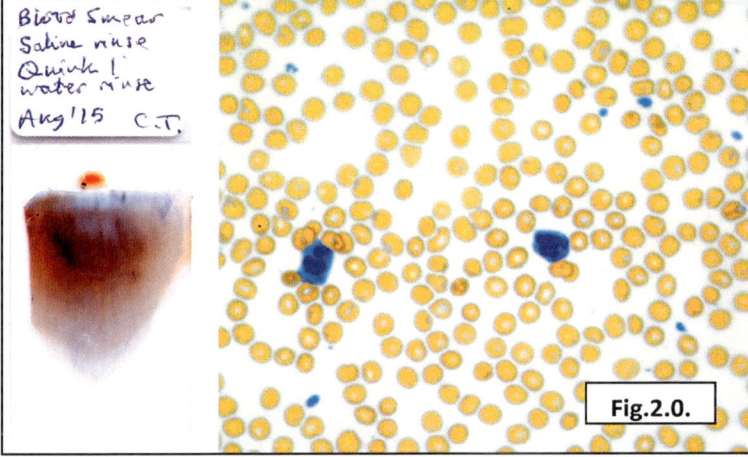

Figure 2. Left - blood smear stained with Quink®. Right - stained blood as seen at 400x under light microscope.

Fig.3.0. (continues next page).

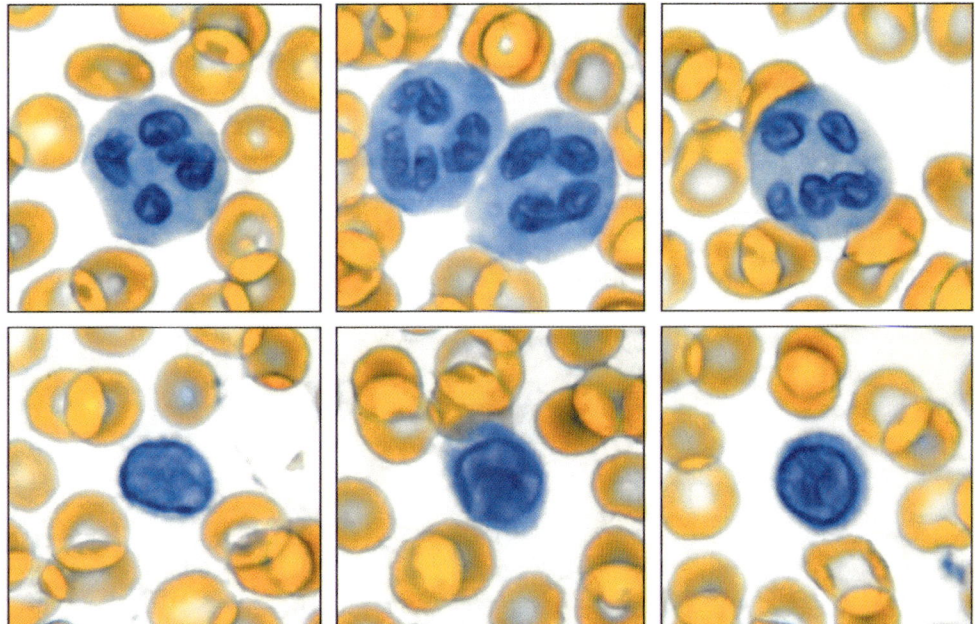

Neutrophils

Small Lymphocytes

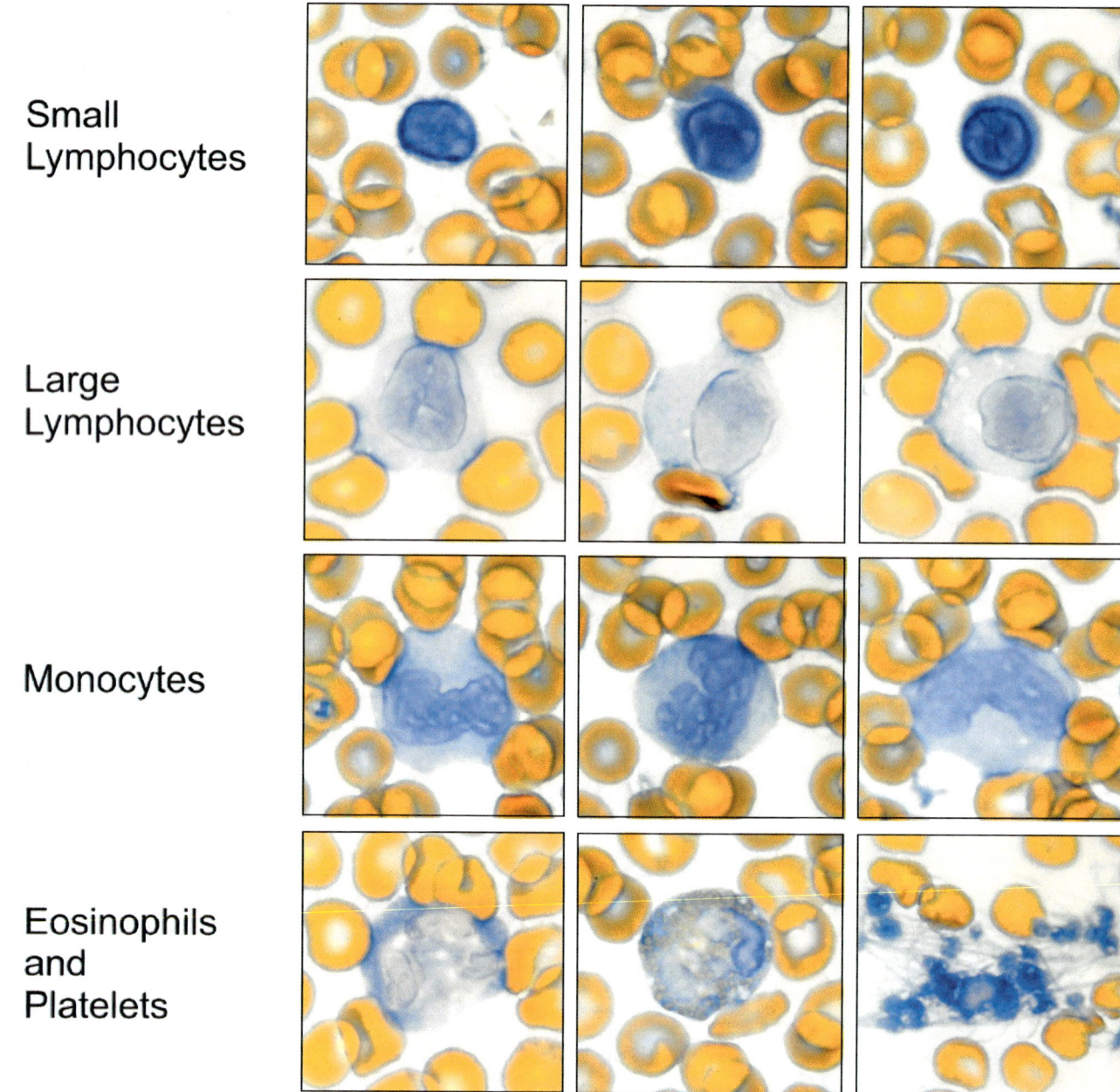

Small Lymphocytes

Large Lymphocytes

Monocytes

Eosinophils and Platelets

Figure 3. Examples of white blood cell types seen after Quink® staining. Viewed at 1000x using oil immersion light microscopy

The cell types are in approximate order of abundance: Erythrocytes (red blood cells—appear orange); platelets (strongly stained blue). neutrophils with multiple lobed nuclei (strong blue staining); small lymphocytes (strong blue staining. Ovoid nucleus, little cytoplasm): large lymphocytes (weaker staining. Larger ovoid nucleus with visible cytoplasm); monocytes with sausage shaped or bilobed nuclei (weaker blue staining): and eosinophils (irregular blue staining and granulation obscuring the cell contents). A rarer type of white blood cell, the basophil, was not seen or identified.

The key element in staining was pre-wetting the fixed blood smear with the 0.8% salt solution. Simply wetting with water also worked, but resulted in uneven staining. Using 0.1% sodium bicarbonate (baking soda, alkali) or ascorbic acid (Vitamin C, acidic) to wet the slides also worked better than water.

Reasonable staining could still be achieved with a twofold dilution of black Quink®.

Staining for 3 minutes gave equally intense staining of the white blood cells. However, the red blood cells took on more stain too, appearing darker.

Staining blood smears with dilutions of 10x or 20x in either water, saline, dilute sodium bicarbonate or vitamin C only gave faint blue staining of white blood cells after 3'. This was the case even after 30 minutes staining (figure 4). The differentiation between red blood cells and white was also lost as the red blood cells became grey blue. If the slide was then restained with undiluted black Quink®, the white blood cells were stained a darker blue (figure 4), as seen with slides originally stained with undiluted Quink®. However, the yellow-orange of the red

*(Below)—**Figure 4.** Left blood smear after staining for 30 minutes with 1:20 dilution of Quink® in saline. Right - Same slide restained using undiluted Quink® for 1 minute.*

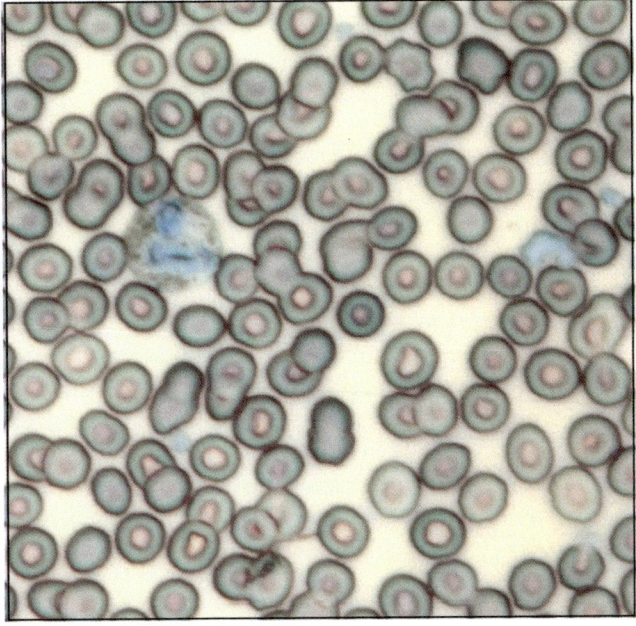
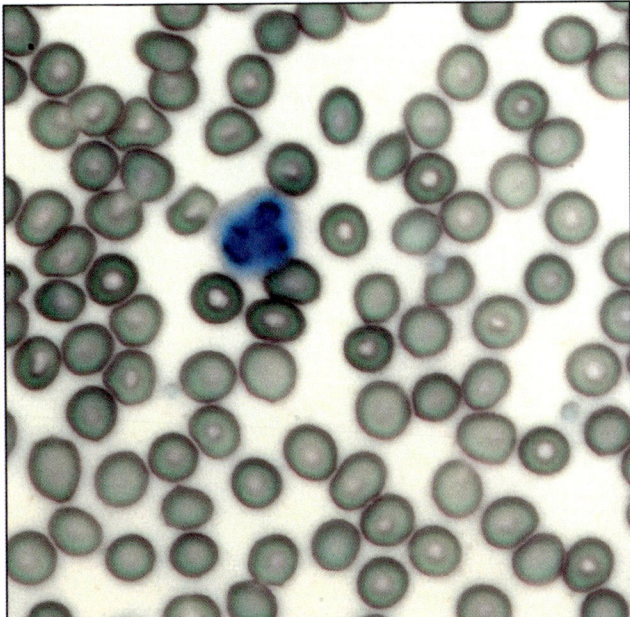

blood cells was not regained.

    Staining worked with a 20+ year old bottle of Parker black Quink® and one purchased in August 2015, the ink seems remarkably stable and consistent (figure 5).

*(Below)—**Figure 5.** Paper towel chromatography of 20+year old Quink® (left) and 2015 Quink® (right). About 0.3ml ink applied to centre of single layer of paper towel, followed by several millilitres of water till orange*

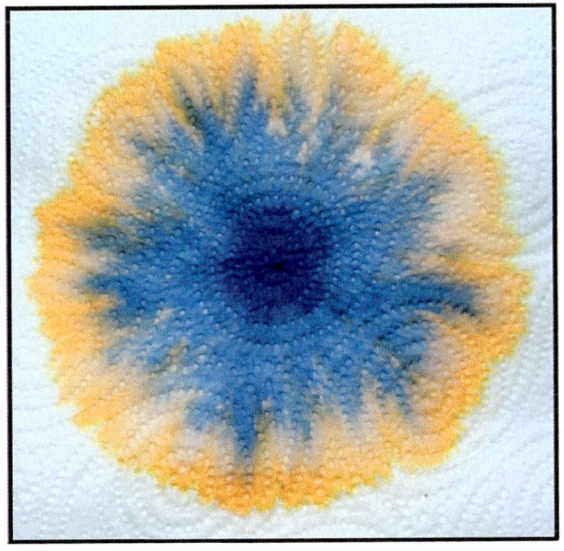
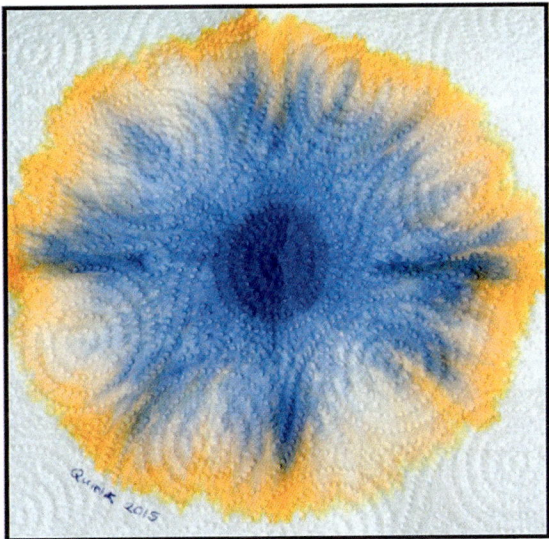

Discussion

This report appears to be the first on the successful use of Quink® as a stain for peripheral blood smears. The stain is simple to use and consistent.

    Quink® appears to make red blood cells (erythrocytes) more visible by weakly binding to the outer surface.

    In white blood cells, the stain penetrates into the cells and preferentially stains nuclear material, with weaker staining of the cytoplasm. Neutrophils, small hymphocytes and platelets stain most intensely. Large lymphocytes, monocytes and eosinophils stain a paler blue. The main benefit of the Quink® stain for peripheral blood smears is for use either:

- In schools or teaching locations where fountain pen ink is seen as a psychologically safer material to other chemical stains
- In places or under conditions where it is difficult to obtain commercial stains
- Where you are found short of your existing stain but do find a bottle of Quink®!

75

Quink® staining is not seen as a replacement for conventional differential blood stains. It is simply another useful tool in the microscopist's workbox.

The current preferred differential stains are Wright's stain, Wright-Giemsa stain and the May-Grünwald stain (1). A faster (15 second) stain, Diff-Quik, is also available. It is based on a modified Romanowsky stain (2). They all stain red blood cells a pale red or similar colour and the white blood cells and platelets in shades of blue or violet. The nuclei are more intensely stained. The Diff-Quik also stains the granules in some white blood cell types red or violet. Platelets are stained blue, violet or purple according to commercial stain used.

Parker's Quink® ink has been used for a number of different staining techniques for more than 50 years. Examples include:

- 1971, Buckley – Fungi pathogenic to man and animals (3)
- 2005, Walker - Fungal mycorrhiza in roots (4)
- 2009, Wikibooks – Staining onion cells (5)
- 2010, Schmitz – Cartilage lesions (6)
- 2012, Mizutani et al – Herpes virus diagnosis (7)

Quink® is a fountain pen ink available in a range of colours. It was developed by Parker as a quick drying ink not requiring the use of blotting paper, using an investment of $68,000 over the three year period 1928-31 (8). The original ink was strongly alkaline and contained isopropanol. A modern Material Safety Data Sheet indicates that it now contains diethylene glycol instead of isopropanol, as well as dyes and preservatives (9).

As can be seen from the chromatograms in figure 5, the black ink is created by combining two dyes of unknown composition, one blue and one a complimentary orange. The dyes are most likely to be derived from aniline salts (10). Other black fountain pen inks that give a similar chromatogram might also work in this procedure.

The Quink® differential staining of blood smears appears to work best using undiluted or twofold diluted Quink®. Mizutani (7) was able to stain the nuclei of herpes giant cells using a 5% solution (twenty fold dilution) of Quink® in phosphate buffered saline. Walker stained fungal mycorrhiza in an acidified 2% (50 fold dilution) of Quink® in dilute acetic acid or HCL (4).

Undiluted Quink® as used in my trials is said to be alkaline. However, staining with a 1:20 dilution of Quink® in either acidic solution, weak alkali or simply 0.8% saline did not give the same strong differential staining as undiluted Quink®. The ability to restain with undiluted Quink® and obtain intensely stained neutrophils suggests that the stainable material within the cells still remains.

The most important other factor for even staining with Quink® was pre-wetting the dried blood smear.

Acidic, alkali or saline washes before staining all seemed to give more consistent staining that wetting with tap water. It was for this reason that I decided that 0.8% salt solution was the simplest pre-wetting liquid.

Post-staining rinsing also appeared to be uninfluenced by either weak acid, alkali or saline solution. Hence the decision to use a water rinse.

Parker's black Quink® ink can be used as a reproducible and reliable stain for peripheral blood smears. It colour differentiates between red and white blood cells. Erythrocytes, neutrophils, large and small lymphocytes, monocytes, eosinophils and platelets can be identified. The stain can be used with generally available ingredients, making it accessible to schools, amateur microscopists and professionals who are either out of stock or unable to obtain conventional laboratory stains.

The method should be transferable for use with blood from other animals. I look forward to your news and results.

Method

Using my own blood samples, this was the final procedure adopted after several trials:

- Wash hand to be sampled with warm soap and water (11)
- Prick the side of a finger near the nail with a sterile needle or blood lancet (11)
- Make an air dried blood smear as described by Schall (12)
- Fix by flooding slide with either alcohol or methylated spirits for 1 minute
- Discard alcohol into waste container
- Allow slide to air dry
- Wet the slide with a few millilitres 0.8% NaCl (salt) dissolved in tap or bottled water. Collect runoff in waste container
- Place slide on flat surface or hold horizontally
- Add 0.5ml to 1.0ml of Parker Black Quink® fountain pen ink to cover wetted blood
- smear. Use undiluted or diluted 1:1 with 0.8% salt solution
- Stain for 1 minute
- Rinse ink off slide with water to into waste container
- Stand slide vertically on tissue paper and allow to air dry
- The stained smear is coloured blue-black to blue grey
- One the slide is dry, it can be viewed under the microscope at 100x and 400x magnification in air or at 100x using oil immersion
- For correct use of a microscope see Understanding and using the light microscope (13)
- Red blood cells appear yellow to dirty orange, in high contrast. White blood cells and platelets are stained blue
- Permanent slides can be made by adding a drop of Canada balsam on the slide and adding a cover slip. Warm the slide to help the Canada balsam spread. Store flat and allow to harden over several days.

Bibliography

1. Wright's Stain. Wikipedia. [Online] [Cited: 18 August 2015.] https://en.wikipedia.org/wiki/Wright%27s_stain

2. Diff-Quik. Wikipedia. [Online] [Cited: 18 August 2015.] https://en.wikipedia.org/wiki/Diff-Quik

3. **Buckley, Helen R.** Fungi pathogenic for man and animals: The subcutaneous and deep-seated mycoses. Methods in Microbiology. 1971, Vol. 4, pp. 461-478.

4. **Walker, Christopher.** A simple blue staining technique for arbuscular mycorrhizal and other root-inhabiting fungi. Inoculum. 2005, Vol. 56, 4, pp. 68-69.

5. School Science/Staining onion cells without methylene blue. Wikibooks. [Online] 2009. [Cited: 16 August 2015.] https://en.wikibooks.org/wiki/School_Science/Staining_onion_cells_without_methylene_blue

6. **Schmitz, N. et al.** Basic methods in histopathology of joint tissues. Osteoarthritis and Cartilage. 2010, Vol. 18, pp. S113-S116.

7. **Mizutani, H. et al.** Single step modified ink staining for Tzanck test: quick detection of herpetic giant cells in Tzanck smear. Journal of Dermatology. 2012, Vol. 39, 2, pp. 138-40.

8. Quink®. Wikipedia. [Online] [Cited: 17 August 2015.] https://en.wikipedia.org/wiki/Quink

9. Material Safety Data Sheet: Parker Quink/Penman Inks . msdsdigital. [Online] Newell Rubbermaid. [Cited: 17 August 2015.] http://msdsdigital.com/system/files/ParkerQuink_96112.pdf

10. **Conner, Rick.** Fountain pen inks. Penspotters. [Online] [Cited: 17 August 2015.] http://www.rickconner.net/penspotters/inks.html

11. **Vann, Madeline.** [Online] [Cited: 15 August 2015.] http://www.everydayhealth.com/diabetes/tips-reduce-finger-prick-pain.aspx

12. **Schall, Jeff.** Making and Staining a Blood Smear. [Online] [Cited: 15 August 2015.] https://www.uvm.edu/~jschall/pdfs/techniques/bloodsmears.pdf

13. **Thomas, Chris.** Understanding and using the light microscope. [Online] [Cited: 15 August 2015.] http://miltoncontact.co.uk/usingthemicroscope

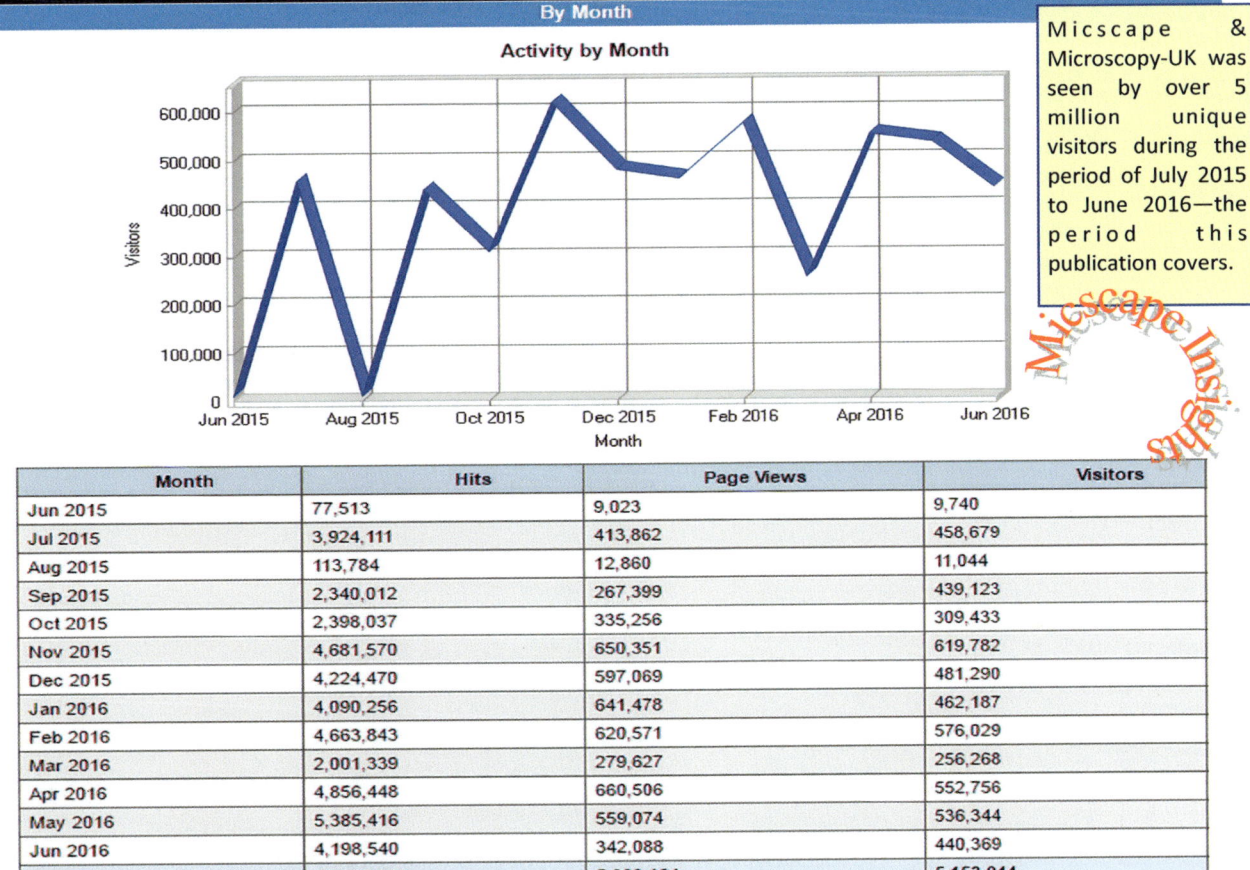

Micscape & Microscopy-UK was seen by over 5 million unique visitors during the period of July 2015 to June 2016—the period this publication covers.

Month	Hits	Page Views	Visitors
Jun 2015	77,513	9,023	9,740
Jul 2015	3,924,111	413,862	458,679
Aug 2015	113,784	12,860	11,044
Sep 2015	2,340,012	267,399	439,123
Oct 2015	2,398,037	335,256	309,433
Nov 2015	4,681,570	650,351	619,782
Dec 2015	4,224,470	597,069	481,290
Jan 2016	4,090,256	641,478	462,187
Feb 2016	4,663,843	620,571	576,029
Mar 2016	2,001,339	279,627	256,268
Apr 2016	4,856,448	660,506	552,756
May 2016	5,385,416	559,074	536,344
Jun 2016	4,198,540	342,088	440,369
Total	**42,955,339**	**5,389,164**	**5,153,044**

WINTER IS FOR MICROSCOPY
II MULTICELLULAR ALGAE
By Anthony Thomas, Canada
Published July 2015

In the April 2015 Issue (of Micscape magazine) I discussed the equipment I use and described some local unicellular freshwater algae. Here in Part 2, I will introduce some multicellular forms.

Haematococcus although a single-celled alga it gives the appearance of being multicellular as it occurs in colonies coating shallow rock pools. It thrives in enriched waters and is a common inhabitant of concrete bird-baths. It occurs in three forms:

> i) a swimming cell consisting of a wide sheath like wall and an an ovoid green chloroplast, often with a spot of orange-red carotenoid pigment; two flagella extend outwards at, what appears to be, the anterior end of the cell (Fig. 1a).

> ii) a round sessile cell, green with some central carotenoid pigment (Fig. 1b).

ii) a round red cyst that is able to withstand drying (Fig. 1c). The red is the strong antioxidant astaxanthin which is thought to protect the resting cyst from the detrimental effects of UV-radiation from direct sunlight when the are in dry conditions.

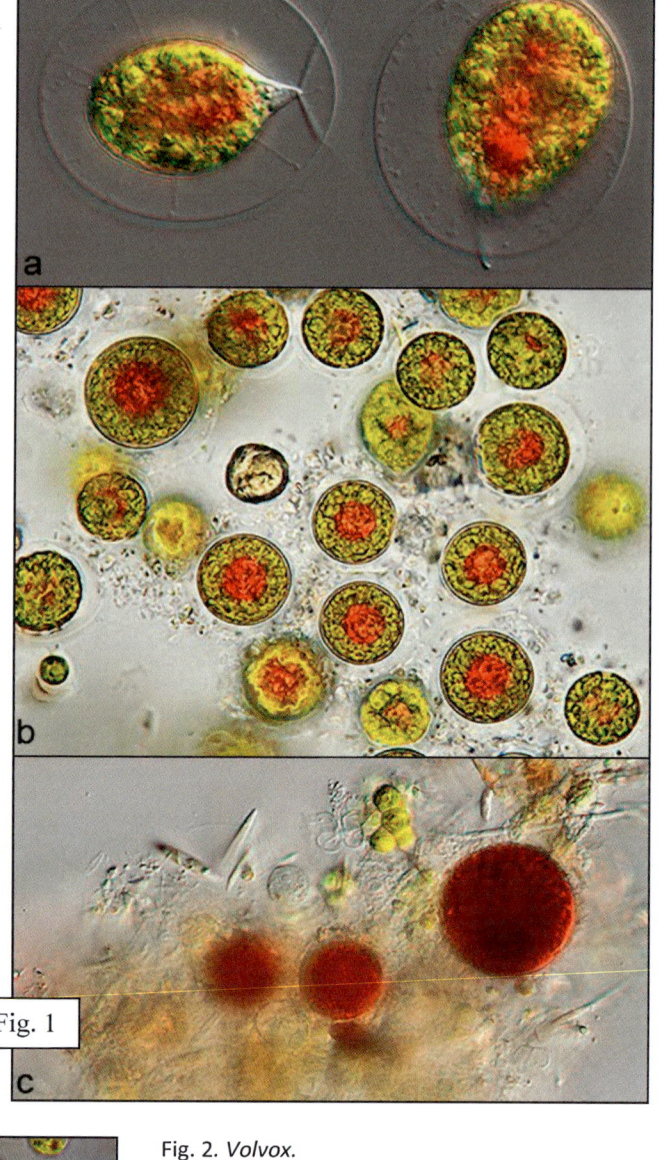

Fig. 1

Fig. 1. *Haematococcus* sp. from my concrete garden birdbath: a: swimming cells, b: normal cells, c: resistant cysts

Volvox is a relatively large, up to 1 mm diameter, actively swimming colonial alga that can contain as many as 50,000 cells. Difficult to photograph owing to its spherical shape and its constantly rotating swimming action. Young colonies are small clear spheres with small green chloroplasts of individual vegetative cells connected to adjoining cells by strands of cytoplasm (Fig. 2a). As the colony matures it produces daughter colonies inside the ball which show up as large green blocks of chlorophyll (Fig. 2b). In late Fall as environmental conditions

Fig. 2

0.1 mm

Fig. 2. *Volvox.*
a: green chloroplasts of individual cells with strands of clear cytoplasm connecting adjacent cells.
b: mature colony showing many small green individual cells and 5 larger daughter cells.

Fig. 3. *Volvox.* Resting spores.

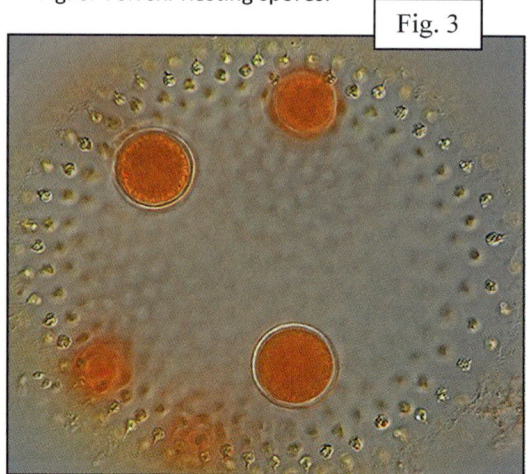

Fig. 3

deteriorate the colony produces resting spores containing the same orange-red pigment seen in the resting spores of *Haematococcus* (Fig. 3). These spores overwinter and start a new colony when conditions improve.

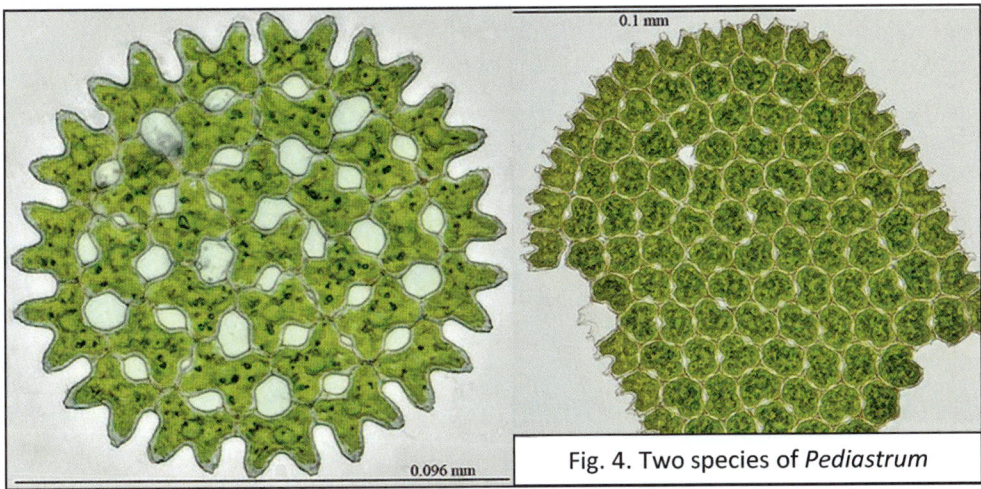

Fig. 4. Two species of *Pediastrum*

Pediastrum species occur as platelike colonies. The few I have seen are circular in outline. In the outer circumference the cells have protrusions/indentations whereas those in the inner plate are more regular. The inner cells may be contiguous (i.e., touching) or there may be spaces between adjacent cells. The plates are somewhat delicate and will break apart if not carefully handled (Fig. 4).

Many species in several genera have individuals living together in sphere of mucilaginous jelly. I think this specimen (Fig. 5) is *Sphaerocystis*. What was one individual has cleaved into several daughter cells (da) and all these are in their own little ball of jelly (je). Note the wide gelatinous sheath (sh) surrounding of the colony. Perhaps the best known, at least in name, unbranched filamentous algae is *Spirogyra*. Each elongate cell (ce) contains one or more spiral chloroplasts (ch), cells are joined end-to-end to make a filament and usually many filaments occur together to form an obvious mat of green algae (Fig. 6).

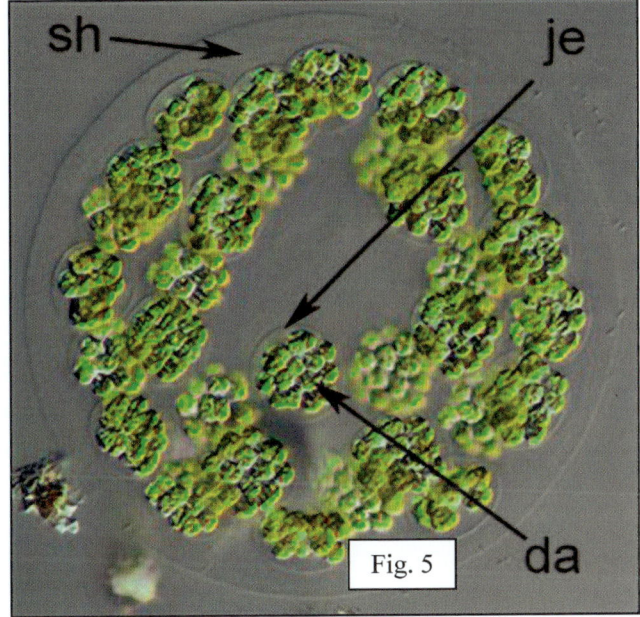

Fig. 6. *Spirogyra*. Top: a typical mass of filaments, each filament a string of individual cells.. Bottom: an individual cell. (*ch*) spiral chloroplast (*ce*) end walls of a single cell.

Above. Fig. 5. Colonial algae in a mucilaginous sphere; possibly *Sphaerocystis*

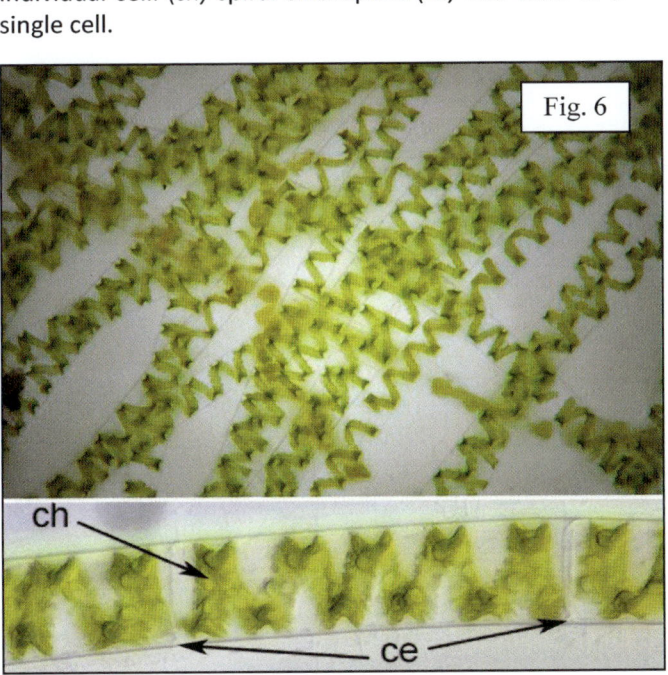

Zygnema is another genus of filamentous algae. Here each individual cell contains two star-shaped chloroplasts which makes identification simple. Some species have the filaments enclosed in a gelatinous sheath, just detectable in my image as fibrils extending outwards from the cell wall (Fig. 7).

Fig. 7. *Zygnema,* showing characteristic chloroplasts

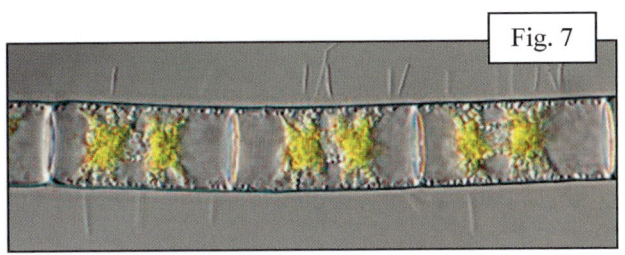

In my text book "Prescott, G.W. 1954. How to know the Freshwater Algae" one of the keys separates the filamentous algae based on the shape of the chloroplasts:

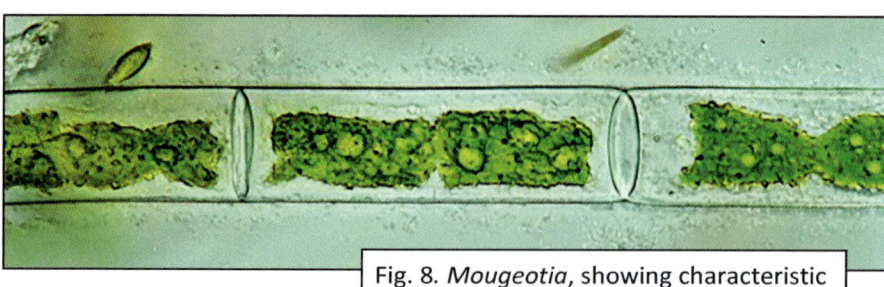

Fig. 8. *Mougeotia*, showing characteristic plate-shaped chloroplasts

a) a parietal ribbon as seen in *Spirogyra*,
b) axial stellate as seen in *Zygnema*, and
c) an axial plate as seen in this genus *Mougeotia* (Fig. 8).

Hyalotheca is another filamentous algae with short, almost square in side view, cells are often filled with green photosynthetic chloroplasts. The entire filament may be encased in thick mucilage sheath as seen in this image, Fig. 9 top, of *H. dissiliens*. In one report I read it was suggested that the sheath contributes to colonial coherence and increases the chance of dispersal as the filaments readily stick to migratory water fowl. I suspect that the sheath would also help keep the filaments moist, when in air, and prevent the algae drying and dying.

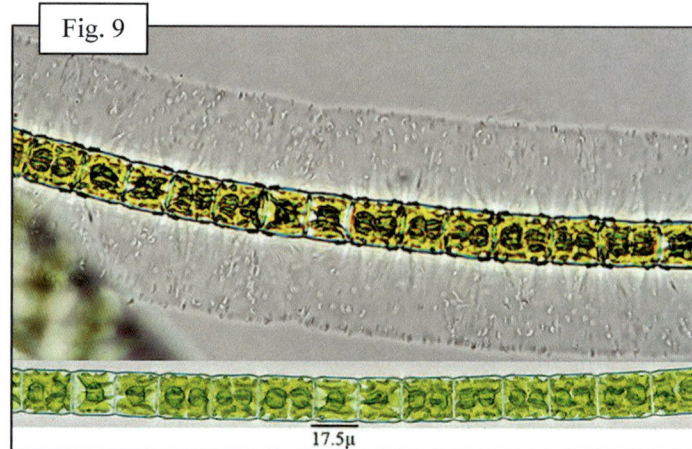

Fig. 9

17.5µ

Besides the filamentous multicellular algae in the previous pages there is a group of multicellular algae that are branched. In this group of branched algae are species in the genus: *Draparnaldia*, consisting of a filament of large cells forming an axis from which tufted plumes of branches of small cells arise (Fig. 10).

Bulbochaete is another branched species but the branches are far fewer (Fig. 11) than the branch structure in *Draparnaldia*.

Bulbochaete can be recognized by the presence of bulb-like bases (Fig. 11, b) of the long 'hairs' projecting from the top of the cells.

100µ

Fig. 10. *Drapanaldia*

Comments to the author are welcomed, email: mothman@nbnet.nb.ca
Article is online is online at: www.microscopy-uk.org.uk/mag/artjul15/at-algae-II.pdf

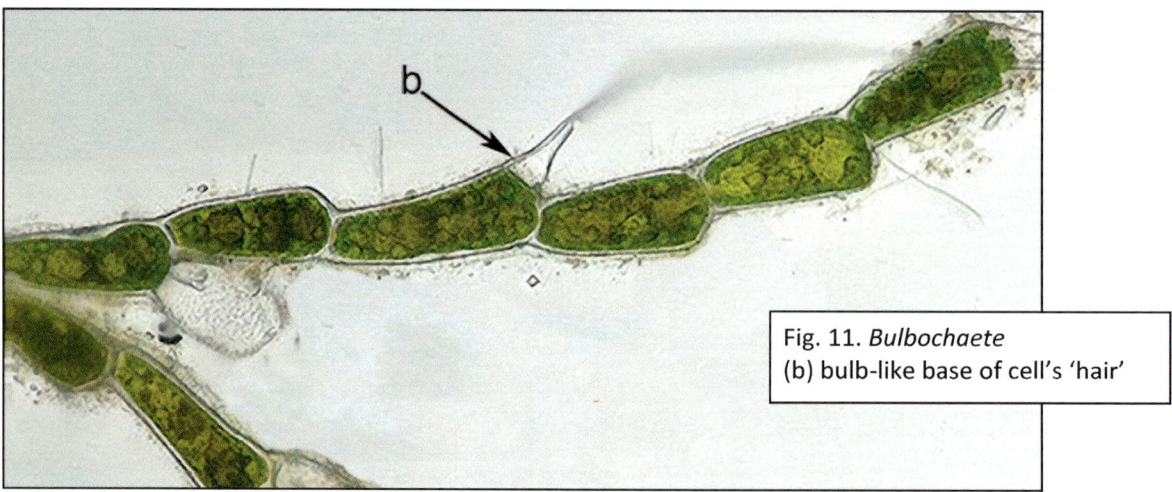

b

Fig. 11. *Bulbochaete*
(b) bulb-like base of cell's 'hair'

The Paramecium Enigma
(Do one cell organisms have a form of intelligence?)

by
Mol Smith
Published November 2014

The world at the microscopic scale perhaps has mysteries which could lead to giant leaps in our understanding of our universe and our reality. One of my main interests is studying human knowledge at the edge of our most advanced levels of understanding and exploration. Much of the leading edge of science has ideas and hypotheses which border on the metaphysical and I believe also leech into other areas of less scientific study - philosophy for example. In fact I am currently writing a book which I hope to publish next year as a result of my own studies. Much of our 'new' gains in scientific application have come about through the study of the quantum world. It is a branch of science (Quantum Physics) where, although our mathematics enable extraordinary exploitation of the effects we encounter in the sub-atomic world, no person on the planet actually has any real grasp of what really happens there. What we discover by experiment defies human intuition and comprehension.

Part of my research has brought to light work by Roger Penrose, an eminent mathematician and one of the bright minds in the early Stephen Hawking camp back a few decades ago. Of particular interest to me is some

Paramecium (c) Richard Howey, USA

of the text from one of his many books 'Shadows of the mind' ISBN 9780099582113. It refers to characteristics of the humble Paramecia (Paramecium *pl.*). This is a single celled organism - an entity contained completely within a single unitary cell. Here is my rough sketch of a type of Paramecium.

A = Macronucleus
D = **Cilia**
G = Gullet
J = Food Vacuole circulating
B = Contractile vacuole

E = Trichocyst
H = Anal pore
C = Radiating canal
F = Micronucleus
I = Food vacuole forming at base of gullet

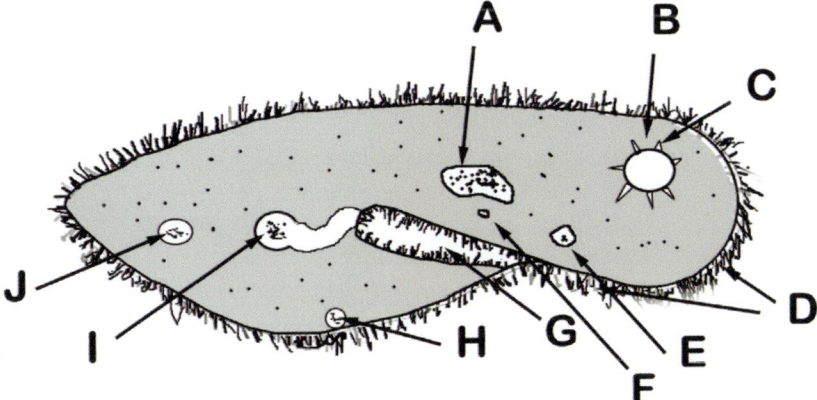

And an image (below) taken at the microscope by Wim van Egmond **(c) Wim van Egmond**

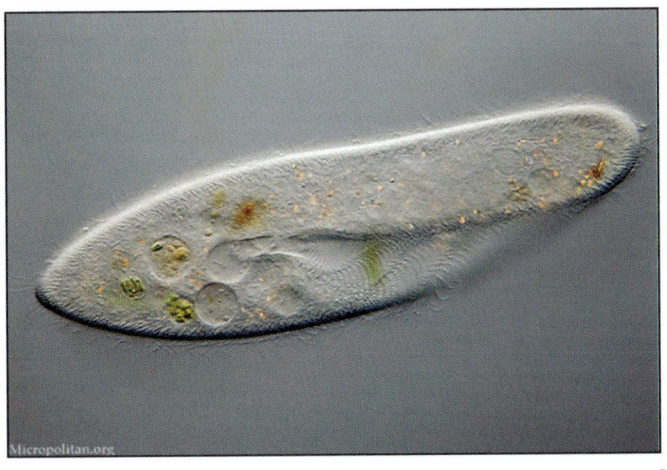

Paramecium possess interesting characteristics and behaviour. They are able to detect food in the form of bacteria, algae, and yeasts in the water, swim towards them by using waving cilia (fine hair-like processes) on the surface of the cell membrane, absorb food into a gullet to create food vacuoles, and they are able to produce by both sexual and non-sexual reproduction, depending on environmental conditions. More... they are also able to detect light, and perhaps the most profound of all - they are able to make a series of manoeuvres to negotiate their way around obstacles they 'bump' up against. In these circumstances, they back up, change their swimming angle, and swim forward again. This might be repeated several times until they find a way past.

Some observers have witnessed behaviours which suggest they are able to retain information in some undefined and inexplicable way, although this suggestion is heatedly challenged by other expert. The enigma though: *what is it that is so mysterious about this tiny life-form?*

Neurons! Yup, nerve cells! They don't have any. How could they? They are, as I said, one celled entities. And yet, a series of actions, preceded by the 'sensing' of an object is apparently orchestrated within the cell to maintain a path towards its objective. Animals in the macro world use nerves and a nervous system to co-ordinate signals from senses to a brain and from brain to muscles to initiate action. Some forms of plants (a well known one being the Venus fly trap) can take simple actions in response to external stimuli. There is no known mechanism for the paramecium to be able to detect (sense) an obstacle, alter the angle of sweep along the cilia to move backwards, time that until it is clear, and change the sweep of the cilia to change the angle of the forward cell motion that then follows. They are considered to be animals, although like many small living forms, such classification can be quite a subjective and difficult conclusion to make. Some single celled organisms behave as plants and also as animals.

If the paramecium had a brain or even a small group of nerve cells somewhere, we could understand a system exists within it to preside over these actions. Nerve cells (neurons) are the only cells we really understand as being able to become an organised bio-mechanism of rudimentary or sophisticated intelligent actions.

It has led Roger Penrose to explore what other processes or mechanisms might be going on as part of his exploration into the human brain and mind. One of the ideas pursued for a number of years is the notion that neurons alone may not be responsible for the kind of profound intelligence shown in human beings, and that there is a possibility that quantum-like activity plays a part in that. Furthermore, for quantum effects to take place, an organ or asset of an organ must be able to resolve or sense quantum effects and in some manner instigate an information exchange to the non-quantum neurons of the brain.

There are processes in the brain which are small enough and abundant enough to be considered potential sites for quantum activity. They are called microtubules. As you will understand, studying anything in the brain, even the activities of brain cells is quite difficult, let alone trying to detect sub-atomic activity at another entirely different scale.

So, here's a thing. **Paramecium have microtubules too!**

Now, let me say straight away, that the paramecium is not the only microscopic life form to exhibit what we would consider to be intelligent activity. Some forms of

Rotifer sense prey so accurately that it's likely they are applying aspects of 'triangular positioning' to pin-point their prey so precisely in both the x and y axes. Once located, they virtually throw a self-contained net, like a hood, (membrane) over the unlucky victim and draw it in. Nearly all the larger microscopic forms exhibit behaviours which seem far more advanced and profound to be the result of chemical-effect reactions per se.

I am going to let Roger Penrose take over for a moment here, by typing a paragraph or two from his book (*without permission being asked for, may I say, but I hope it might encourage other people to buy it*)...

There must indeed be a complicated control system governing the behaviour of a paramecium - or indeed other one-celled animals like the amoeba - but it is *not* a nervous system. The system responsible is apparently part of what is referred to as the *cytoskeleton*. As its name suggest, the cytoskeleton provides the framework that holds the cell in shape, but it does more. The cilia themselves are endings of the cytoskeleton fibres, but the cytoskeleton also seems also to contain the control system for the cell, in addition to providing 'conveyor belts' for the transporting of various molecules from one place to another. In short, the cytoskeleton appears to play a role for the single cell, rather like a combination of skeleton, muscle systems, legs, blood circulatory system, and nervous system all rolled into one!

It is the cytoskeleton's role as the cell's 'nervous system' that will have the importance here. For our own neurons are themselves single cells, and each neuron has its *own* cytoskeleton! Does this mean that there is a sense in which each neuron might itself have something akin to its own 'personal nervous system'? This is an intriguing issue, and a number of scientists have been coming round to the view that something of this general nature might actually be true!

———————————

I'll leave Roger there but he does quote references to publications which support his final statement. His book goes on to closely examine those cilia and their microtubule make-up with quite staggering implications on the possibility that the human brain interacts with the quantum world directly.

But I wanted to introduce the enigma, and its done. I do myself like the idea that study of other living forms and the way they negotiate their external reality, especially with micro-organisms which have been here a lot longer than other life forms, might yield important and hitherto-missed information and clues about the workings of macro life forms. In this case... us.

Mol. E: molsmith@fastmail.fm
Article is on line at:
www.microscopy-uk.org.uk/mag/artnov14/paramecium-engima.html

Notes From The Editor

As this is the first printed version of the Micscape online magazine, and in case for some reason it is the only paper document ever published of so much endeavour by so many people, I decided to 'guest' many contributors work here. This is despite the fact their work may not have appeared in the period this journal covers. Often, many contributors offer material of high quality over a long period of time. And sadly, some authors who contributed over the last 21 years have passed away.

The names of all contributors are too many to list here but a list is maintained online by David Walker and can be viewed at:

www.microscopy-uk.org.uk/mag/authors.html

The table to the right records the number of contributors from each country since the start of Micscape up to and including June 2016.

Apart from the magazine itself, the web presence which is Microscopy-UK that hosts Micscape is an integral part of the suite of resources provided for enthusiast microscopists. Several major sections are maintained. These are as follow:

The Micropolitan Museum which is divided into smaller sections...
>	The Freshwater Collection
>	The Marine Collection
>	The Insectarium
>	The Botanical Garden

And is created and maintained by Wim van Egmond in The Netherlands.
www.microscopy-uk.org.uk/micropolitan/ x_index.html

A Flower Garden of Macroscopic Delights (and other occasional non-flowering plants) created by Brian Johnston in Canada.
www.microscopy-uk.org.uk/mag/bj-flowers.html

The Colourful World of Chemical Crystals created by Brian Johnston (Canada)
www.microscopy-uk.org.uk/mag/bj-crystals.html

The Pond Life Identification Kit created and maintained by Wim van Egmond in The Netherlands.
www.microscopy-uk.org.uk/pond/index.html

The Online 3D Microscope created and maintained by Mol Smith.
http://www.microscopy-uk.org.uk/3d-microscope/ index.htm

An Online Library of all past Micscape articles maintained by David Walker.
www.microscopy-uk.org.uk/mag/libindex.html

Country	Number of contributors
Australia	13
Austria	2
Belgium	3
Brazil	3
Bulgaria	1
Canada	10
China	2
Estonia	2
Finland	2
France	5
Germany	4
Greece	2
Israel	1
Italy	4
Japan	2
Jersey	1
Mexico	2
N. Ireland	1
Netherlands	15
New Zealand	5
Norway	2
Philippines	1
Puerto Rico	1
Russia	4
South Africa	1
Spain	1
Thailand	1
UK	79
USA (includes 131 Rochester Institute of Technology, NY students' course contributions.)	200

'Abridged.'

A Close-up View of Two "Parrot Tulips"
Tulipa x hybride

Part 1
Published November 2014
by Brian Johnston (Canada)
www.microscopy-uk.org.uk/mag/artmay14/hc-flowers.html

The flamboyant flowers discussed in the two articles this month are certainly different than the normal garden-variety tulips found in most spring gardens. What makes them so special to a macro-photographer, is their continued visual interest as one moves up the magnification scale. Each bud can be considered to be a colourful sculpture, with striking structure that continues to interest the observer as he or she moves closer.

Although tulips are often associated with the Netherlands, they are not a native Dutch flower! About four hundred years ago Europeans first discovered tulips in Turkey. At that time Carolus Clusius, a famous botanist, introduced the plant to the Leiden botanical gardens in Holland. Since tulips were extremely rare, and expensive, only Kings and Emperors could afford to plant them in their gardens.

The immediate popularity of the tulip drove Clusius and other horticulturalists to produce new colour variations to satisfy the growing demand for the flowers. Over the years, many tulip forms were produced by crossing and hybridizing techniques. Some had frilly petals, and dramatic flame-like colourations, that later became known as "Parrot tulips". In the 20th century, these distinctive characteristics were found to be the symptoms of the **mosaic virus** which was transported to the tulip plant by a louse living on peaches and potatoes! Today, hybrids have been developed with similar visual characteristics, but without the virus infection.

In your mind, compare normal tulips, with their single-coloured, solid, smooth petals, with the parrot variety that can be seen above, and in the two images that follow. Parrot tulips are characterized by petals that are curled and twisted. They also have borders that are fringed (**laciniate**). Both varieties share the same characteristic lance-shaped (**lanceolate**), bright green leaves.

Notice the interesting, rather random colouring of individual petals in the two images (*right*). The colour ranges from bright red, through yellow, to almost black

Petals have an irregular, three-dimensional fringe along their upper edge, as well as

occasional projections of various sizes that emanate from their outer edges.

Viewed from behind, the smooth green stem supports the base of the flower. In fact, the stem is the weak point of many parrot tulip varieties, as it is often too weak to support the fully open flower. For this reason, many parrot tulips are suitable only as cut flowers.

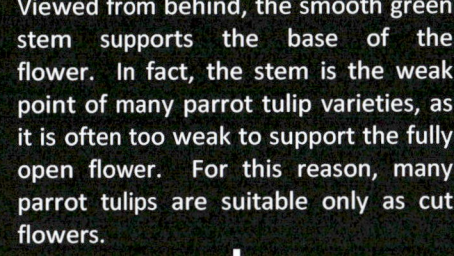

Just how flamboyant the blooms are, can be seen below. Notice that unlike normal tulip petals, each parrot tulip petal has its own individual characteristics. Rotating a bloom shows a different structure with each viewpoint.

Each flower has six petals arranged with three in an inner circle, and three in an outer circle. As a flower begins to open, the outer three petals move away first.

As we move closer to the petals of a flower, the sculptural nature of the surfaces becomes more evident.

Closer still, the petals resemble the wildest imaginings of a hallucinating modern sculptor!

If a section of one of the petals is examined under the microscope using increasing magnification, its cellular structure eventually becomes visible (*below*).

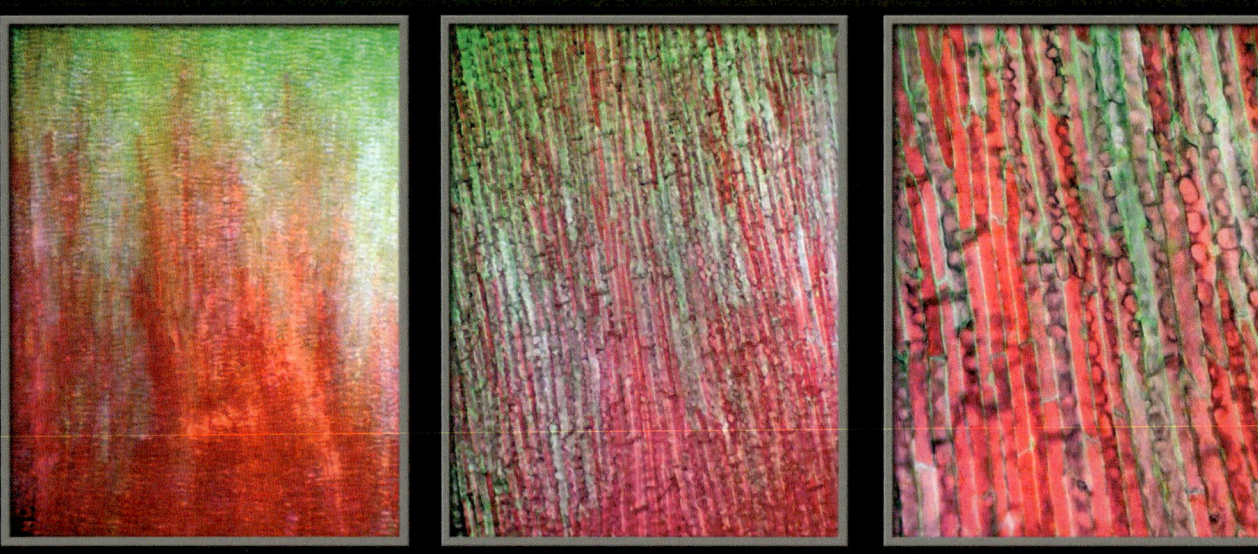

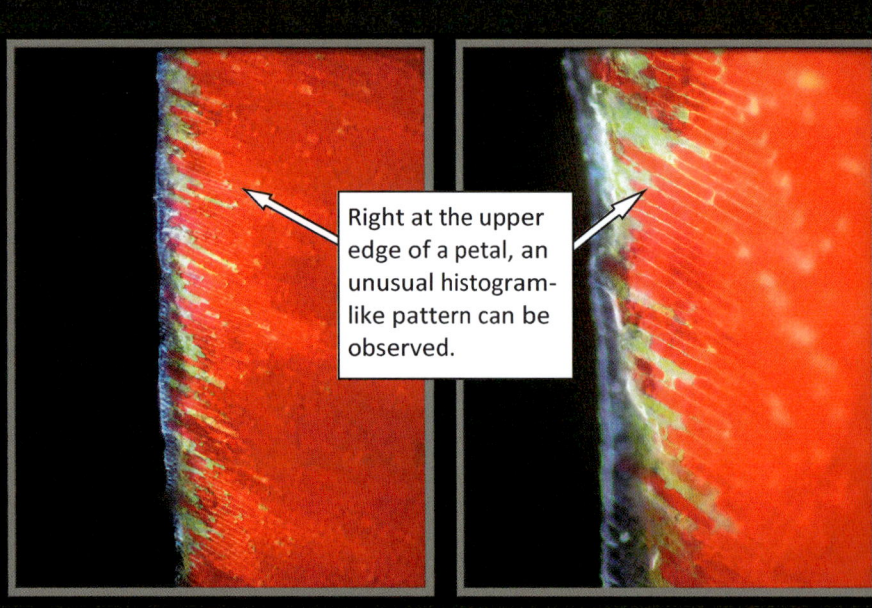

Right at the upper edge of a petal, an unusual histogram-like pattern can be observed.

When a flower opens sufficiently, its reproductive structures become visible

petals. Protruding from the top of the stalk is the white pistil, topped by the stigma. Surrounding the pistil is a ring of six stamens.

greenish-white ovary. A stigma without a stalk (or style) is referred to as **sessile.** The six pollen covered anthers are supported by rod-like filaments.

Notice in the more highly magnified view of an anther shown *right* that it is composed of four parallel segments. Tulip pollen is dark in colour and can be seen clearly if it gets on hands or light-coloured clothing. (The dark out-of-focus spots that can be seen on the yellow base of the petal at left are actually clumps of this dark pollen).

When examined under the microscope, the cellular structure of one of the supporting filaments is colourful! (*Below*).

Several clumps of irregularly shaped pollen grains can be seen clinging to a filament in the two images *right*.

Several anthers in the flower appeared to have this strange folded shape at their bases. At this magnification it is almost possible to see individual pollen grains on the anther's surface (*below*).

By sliding the dark-ground condenser slightly out of the optical path, it is possible to show the pollen grains adhering to the edge of an anther. In this light, they appear to have a purple colour (*below*).

This same purple colouration also appears in a higher magnification phase-contrast view of a single pollen grain.

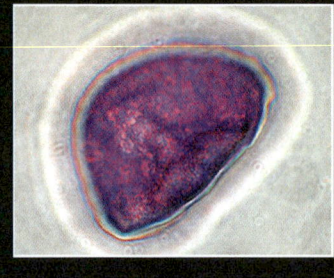

The three images below show the top of the tri-lobed stigma. Each lobe appears to have a deep groove in its upper surface, no doubt to increase its surface area (and therefore its pollen collecting capability). Dark pollen grains can be seen clinging to the stigma's surface in all three images.

Photomicrographs showing the surface of the stigma can be seen right and below. Note the hair-like protuberances that cover the surface. Pollen grains, carried to the stigma by visiting insects, are caught by these hairs.

"Botanical tulips" like the parrot tulips studied in this article, are marvels of what modern biological science can produce. How strange it is to consider that a louse carried virus produced such stunningly colourful blooms. How wonderful that science could retain the striking characteristics, while removing the disease! In part two of this article, a different parrot tulip, with even stranger sculptural qualities, is examined macro-photographically.

Photographic Equipment

The macro-photographs were taken with an eight megapixel Canon 20D DSLR equipped with a Canon EF 100 mm f 2.8 Macro lens which focuses to 1:1. A Canon 250D achromatic close-up lens was used to obtain higher magnifications in several images.

The photomicrographs were taken with a Leitz SM-Pol microscope (using dark ground and phase-contrast condensers), and the Coolpix 4500.

Further Information

Tulips
http://bell.lib.umn.edu/Products/tulips.html

The history of the Tulip
http://www.geocities.com/Salhanie/contents.html

Tulips

http://www.holland.nl/uk/holland/sights/tulips-history.html
Tulip (Tulipa)
http://encyclopedia.jrank.org/TOO_TUM/TULIP_Tulipa_.html

How did a flower cause an economic disaster?
http://www.killerplants.com/plants-that-changed-history/20020402.asp

Editor Introduction. Walter Dioni contributed for many years from his home in Mexico. His articles are both fascinating and helpful. He explored new ways to preserve, mount, and stain materials for microscopical study. A collection of such articles were published by us in a book **Safe Microscopy Techniques by Walter Dioni** which is available from Amazon. Sadly, Walter passed away. But his legacy is profound. I introduce two articles from Walter which were published in Micscape a number of years ago. www.microscopy-uk.org.uk/mag/artapr04/wdslicera.html

A CHEAP AND PRECISE SLICER FOR TEACHING BOTANY
(and new adventures in my garden)
By WALTER DIONI   Durango (Dgo) Mexico
www.microscopy-uk.org.uk/mag/artapr04/wdslicera.html

Introduction

After my <u>last article</u> (*www.microscopy-uk.org.uk/mag/ artfeb04/wdstem.html*), I searched the Web for "double edged razor blades". It was a surprise to me that 10 pages (at least 200 articles) were offered for my search. Many technicians, in several branches of science, do use or want to use these antiquities. There are even collectors that have spent decades collecting blades. The most interesting thing to me is the fact that you can buy the blades in the southern hemisphere and that a few producers in the north sell a limited quantity. Apart from Gillette, there are BIC, Schick, and America Safety Razors; all four have production plants in France, the EU and even Mexico. For those that are near a Wal-Mart supermarket there is an article on the Web stating that they sell our precious blades.

So I have combed Cancún for old two-edged razor blades, and finally I get plenty of Permasharp ones, made in Brazil.

Of course I immediately indulged in new adventures with the Neuburg slicer. (I prefer "slicer" to "microtome" because the sections are not really thin enough to warrant the more technical name.)

After several trials I became convinced that with this instrument the thickness of the sections was unpredictable. The blades are flexible. My fingers always apply different amounts of pressure. And so on.

As a consequence successes are less than failures. So I have finished making 3 additions and 1 major modification (with an additional option) to the little instrument.

1. I prepared cutting surfaces, making 1.5 cm wide and 3 mm thick strips from the expanded polystyrene trays my
2. I buy a paper clamp (see fig 1) 32 mm wide (40 mm could be better).
3. I get a tray of 9 x 20 x 2.5 cm (also from the supermarket) to be filled up with water.
4. In my later attempts I have changed to "Scotch" adhesive tape to separate the blades. The tape has a thickness of nearly 50 microns. I put one piece along each of the lateral sides of one of the razor blades.
5. For the reasons I explain in the <u>second part</u> of this article I now use, as an alternative, half a razor blade for the job. The thickness of the blades is nearly 100

microns. I put a half broken razor blade along the upper edge of the slicer, without using the tape at all. See the technical tips.

When I want to make a section I put the blades together and clamp them with the jaws of the clip just over the slits so they are well fastened. Submerge the edges in the water, or put some drops over one blade before closing the slicer. Water goes up by capillarity between the blades.

Now I put the biological material over the cutting strip, press it with my finger, taking it all in my left hand and submerge the whole under the water in the cutting tray.

With the other hand I place the cutting area of the blades in position over the material. Pressing the instrument down and ahead with a diagonal trajectory I cut slowly until both edges indent the cutting surface. This is important because this ensures that the section is completely separated from the cut material. A plastic strip 10 cm long supports dozens of attempts.

Now I remove the clip and, best under water, manually or with the aid of the point of a needle or a scalpel I open the razor blades.

Normally there is a beautiful section stuck near the cutting edge of one of the blades. With a fine and soft brush I pass it to a two cm deep capsule with a few millilitres of 50% glycerin .

To cut another section, I rebuild the razor blades

medium thick section, x 10 objective, 9 unit pictures

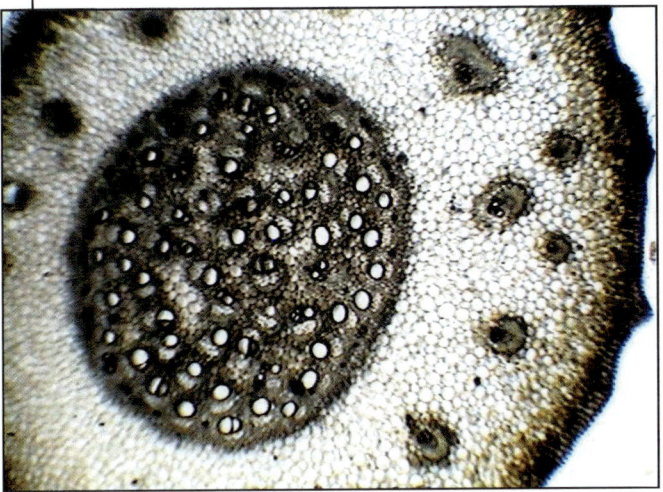

sandwich and press it with the clamp.

Above: thick section, x 4 objective
all three pictures in bright field illumination
all three sections from Epipremnum aureum

In a few minutes one can make several subsequent sections, obtaining consistent results. Used over the polystyrene surface the sharpness of the blade edges are maintained for many hours (even days). You need of course to find the proper position of the clip's jaws over (or under) the razor slits. More than any other feature this governs the thickness of the section.

Try putting the jaws just over or not more than 4 mm from the upper edge of the blades' slits. In my experience within these limits a higher setting gives thinner sections. Isn't it amazing?

The most common sections are transversal ones (cross sections, x-sections) but with this new configuration you can make longitudinal sections. Use a reasonably wide stem ca. 5-6 mm long. Take it between

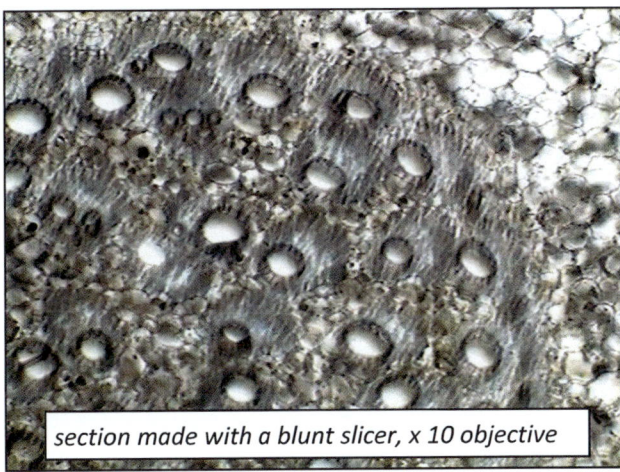

section made with a blunt slicer, x 10 objective

your fingers and slide the slicer carefully between them to reach the surface of the stem. Make the section taking care the blades are vertical. Not all attempts are successful. I normally make 3 to 5 attempts to finally select a thin, transparent and uniform section.

Below: a composite image of a thin vertical section of an Epipremnum stem. Bright field, x 10 objective. Five individual pictures stitched with Photopaint.

You can mount your sections temporarily in water, or in 50% glycerol in water. The latter has a very good refractive index and lasts several hours with minimal replenishment. You can even use a little Vaseline on the coverslip borders as is customary for wet mounts.

You know your sections are thin enough when the coverslip lays flat over it, neither tilting nor requiring many drops of media.

Sometimes I work in small batches.

The capsule with the sections gathered after one working session is put on the turntable of my microwave oven and treated with 12 to 15 seconds of irradiation. I mount them in pure glycerin or in PVA-G, and set them aside for at least 12 hours before bringing them to the microscope.

Alternatively I collect the sections in water, and pass them to hypochlorite solution (6% free chlorine) if I intend to make a more permanent slide.

Permanent slides could be made with 60% fructose, GAF, PVA-G or glycerin jelly, without any subsequent treatment but an adequate stepping in the concentrations of the mounting media (or glycerin infiltration, in the glycerin jelly case). I have even mounted directly in PVA-G. These slides are good for a general view of the histology of the studied organs.

And with both alternatives you can profit from an assortment of contrast discs** which gives you the benefit of diffraction staining, besides the normal brightfield illumination.

After the drawing or photographic session comes to an end, the sections can be recovered, washed in distilled water, and submitted to a more classical and permanent mounting technique.

The classical treatment requires getting rid of the cytoplasm with hypochlorite and the differential staining of the remaining cell walls with one or more dyes.

For the construction and use of contrast discs, and the optical equipment used, see Dioni:
www.microscopy-uk.org.uk/mag/artdec03/ wdonion2.html

To learn to make permanent mounts the classical style, see these web references:

http://www2.ac-lyon.fr/enseigne/biologie/ress/ biologie_vegetale/cou_veg.html

http://www.ualr.edu/~botany/celltiss_lab.html

http://www.zoo.utoronto.ca/able/volumes/vol-19/09-yeung/09-yeung.htm

A very good technical paper in two parts covering state of the art techniques for making and mounting botanical sections is presented by Jim Battersby in the 2004 February and March editions of Micscape Magazine.

In a companion article (*www.microscopy-uk.org.uk/ mag/artapr04/wdslicerb.html*), I gather the technical tips for the slicer design and many illustrations of the performance of the slicer.

Pros

1. Doesn't need tissues support (Elder pith, polystyrene, carrot, potato, paraffin wax) which is by itself a huge achievement. Think on this because it is an outstanding feature. Most of the amateurs' discussions on the Web are over the support for tissues to be cut with the Ranvier style microtomes.

2. Section quality is sufficient for a detailed anatomical study of stems, petioles, ovaries of many flowers, leafs, and so on. Leaves are dealt with easily with the new configuration. They are difficult materials for the traditional hand microtomes, not to speak of the essays discussing how to make free-hand sections of them. The problem is, that for a section laying on its cut side its width must be thinner than the thickness of the foliar lamina. The new configuration ensures this. Of course if you work with such thin sections (both in height and width) you can expect some mis-manipulations leading to a twisted lamina, but you always have enough spare material to study the leaf

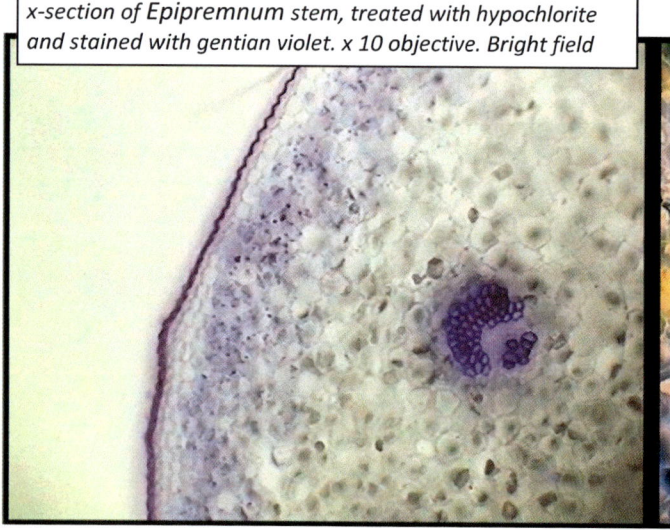

x-section of Epipremnum stem, treated with hypochlorite and stained with gentian violet. x 10 objective. Bright field

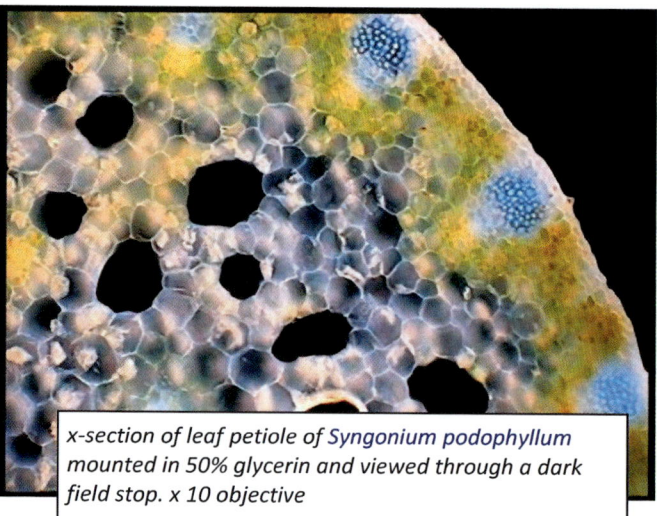

x-section of leaf petiole of *Syngonium podophyllum* mounted in 50% glycerin and viewed through a dark field stop. x 10 objective

anatomy in cross section.

3. Very cheap. Five instruments require two boxes of razor blades (5 razors a box). With a cost (now at Cancún) of 0.28 dollars (0.056 dollars/slicer).

Below: moulds over the epidermis of a x-section of Gerbera leaf. Three focus levels combined with CombineZ. x 40 objective. Bright field

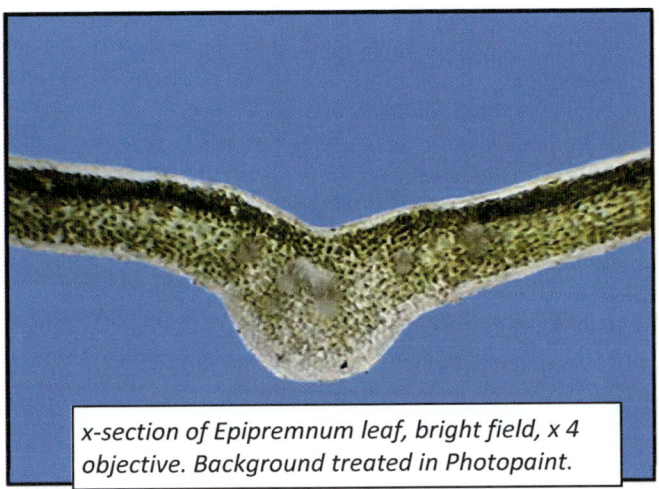

x-section of Epipremnum leaf, bright field, x 4 objective. Background treated in Photopaint.

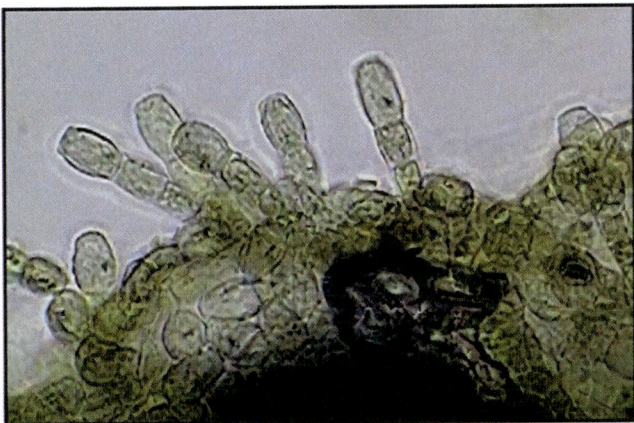

4. Easy construction by careful experienced amateurs. Not more than 10 minutes needed to make and put to work a new slicer using the half razor blade separator, or more or less 20 minutes for the Scotch tape version.

5. Easy to use. The learning curve is very quick. Any user can start to do a good job in a matter of minutes.

6. Safe enough to be used under adult supervision even by secondary school students.

Right: thick section (to conserve the seeds in it) of a mature ovary (or young fruit) of Hybiscus rosa-sinensis. This picture is a mosaic of ten individual ones stitched with PhotoPaint.

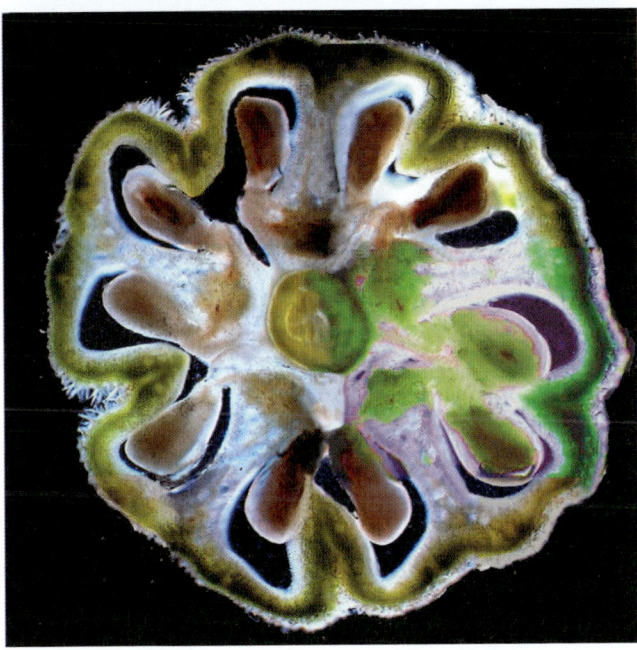

Cons

1. **Not safe enough** to be built by very young scholars or amateurs, without adult supervision. *Of course no one microtome is safe and all professional or even those amateurs' commercial ones are MORE dangerous, and only appropriate for use by technically trained adults.*

2. **Sections must be made one by one.** You need to put together all parts, to make the section very carefully, to split open the instrument, and to pass the sections on to its further destiny. And repeat all of this for any one section. But are you very pressed? Did you need a lot of serial sections in a limited time? Don't try to make several subsequent sections without taking out the first one. The thickness of the first section opens the blades and every new section is wider than the previous one.

3. Air bubbles. It is easy to trap air bubbles in the cut exposed cells or vessels, if you cut dry in the air. Cutting under water mostly obviates this. If there's some persistent bubbles, put the sections in a glass capsule (a little Petri dish is best) in more or less 10 ml of 50% glycerin and submit them to the microwave oven. In a 700W domestic one, at 100% (Full) 10 or 12 seconds, get rid of them. (Make proportional estimates for 400, 600, or 1000W ovens). Additionally the microwave generated heat fixes the plant tissues and clarifies the sections. A word of warning: the Euphorbiacea and many other plants can have lacticiferous channels full of latex that flows out like milk over the cut surface. Cutting under water and removing after some seconds the just made section, generally washes out the latex.

Conclusions

With this new configuration and including the easy to build and cheap contrast discs, and one or two dyes easily found in drugstores, the double razor blade slicer merits incorporation into not only the amateurs' laboratories, but also to the secondary or even more advanced school laboratories.

Its simple construction and use, with consistent good section quality, puts in the hands of every teacher the capacity to give his (or her) students a basic appraisal of the vegetable's anatomy at a more than reasonably low cost. It is really a better choice than the potentially dangerous free-hand cutting method, the original Neuberg design, or all the many recommended elementary, home-made, low tech, bolt and nut based "microtomes"

Gallery

I present here the stars of my story. You will be introduced to more of their secrets and those of other garden plants in the second half of this article.

Ocinum basilicum - basil

A monocotyledon plant - *Syngonium podophyllum*

One domesticated garden cultivar of *Epipremmnum aureum,* a tropical vine.

The flower of *Hybiscus rosa-sinensis.* It was its young fruit that was sectioned.

Plant pictures taken at 1280 x 960 pxs with a Samsung Digimax 201 camera and reduced to 320 x 240 to be inserted here. The foliage under the Hybiscus flower is Aptenia cordiflora to be treated later.

**I prefer to say "contrast discs" because it is shorter than "stops, diaphragms, and filters". After all they are discs, or are mounted on discs, to be put in the filter tray under the condenser of the microscope.

Editor.
Walter wrote so many articles of genuine value through a long period (more than a decade) through ailing health (blindness), ageing biology, and with quite meagre means.
He deserves not one book but many and in time, that may happen. Meanwhile, if you browse this link:
www.mic-uk.info
and enter the search term 'Walter Dioni', you will be able to see the wonderful material he shared with the world.

TECHNICAL TIPS ON THE USE OF THE PRECISE DOUBLE RAZOR BLADES SLICER

Published April 2004

By WALTER DIONI   Durango (Dgo) Mexico

www.microscopy-uk.org.uk/mag/artapr04/wdslicerb.html

I bring together here several tips about the design and performance of the slicer described in the first article (www.microscopy-uk.org.uk/mag/artapr04/wdslicera.html). I add only the minimum technical information for the pictures, because in successive articles I hope to present short illustrated monographs of some plant species, studied with the aid of the slicer. A formulary with laboratory protocols is also a future project.

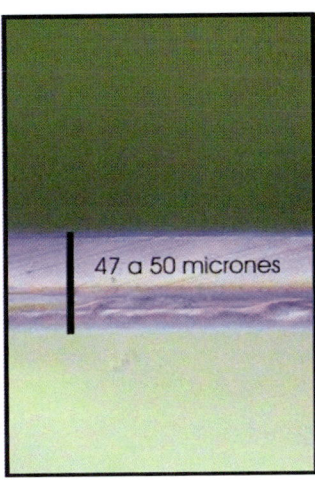

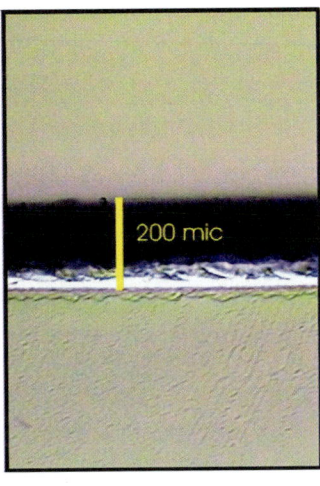

Adhesive tape, x40   **Insulation tape, x10**

On the nature of the separators.- I have tried four kinds of separators. The self-adhesive paper labels, one black plastic electrical insulation tape, one brand of clear adhesive so-called "Scotch" tape and the razor blade itself. To determine their thickness I stick a sample of the tapes to the edge of a glass slide, and cut it level with a razor blade. I take pictures with the 10 x and 40 x objectives using the COL-D3 contrast disc that gives a very good optical separation of the components, and I measure the thickness with the calibrated measuring tool of the camera program. The razor blade was broken into two halves and one half was shaped as a V that can be put upright over a glass slide to offer the sharp edge to the objectives.

The results are confusing, not about the materials but about the thickness of the cut materials.

The limits for the thickness of the sections.- The Scotch tape (including the relatively wide layer of adhesive is 47-48 (roughly 50) microns thick. The adhesive paper label is 135 microns thick; the sharp edge insulator tape is roughly 200. And the thickness of the razor blade itself turns out to be 100 microns.

I make one more measurement: the width of the cutting edge (not its thickness). It turned out to be 300 microns wide. So the sharp edge is an isosceles triangle with a base of 100 microns, and a height of 300 microns. The edge is at 50 microns from each side of the blade.

Using the Scotch tape as a separator I make the razor blade sandwich and put it upright under the 4 x and the 10 x objectives. The width of the gap between the blades is roughly 50 microns as it can be suspected.

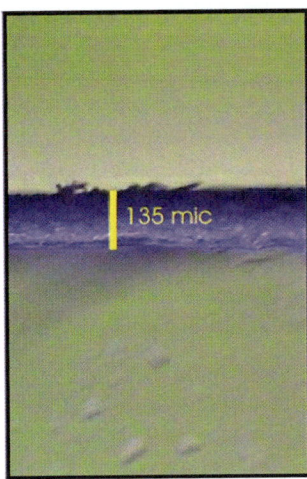

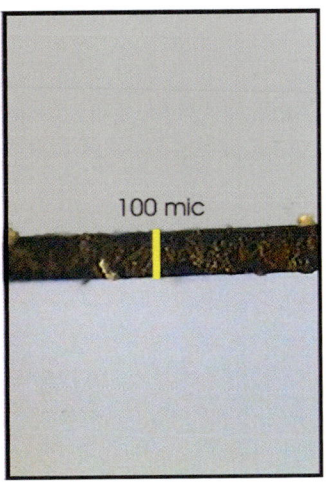

Paper label, x10   **Razor blade x10**

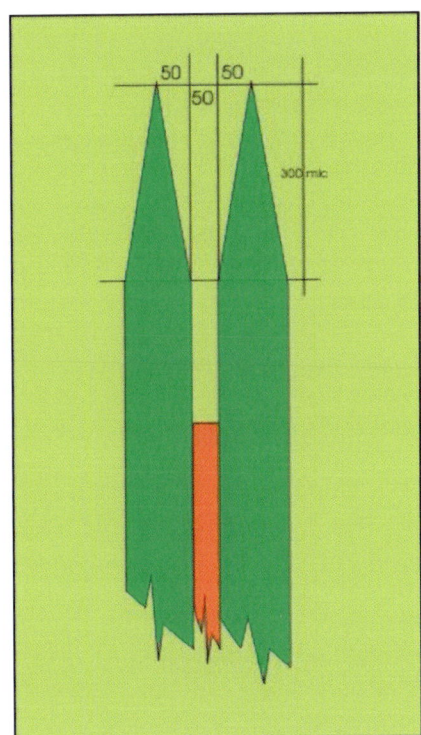

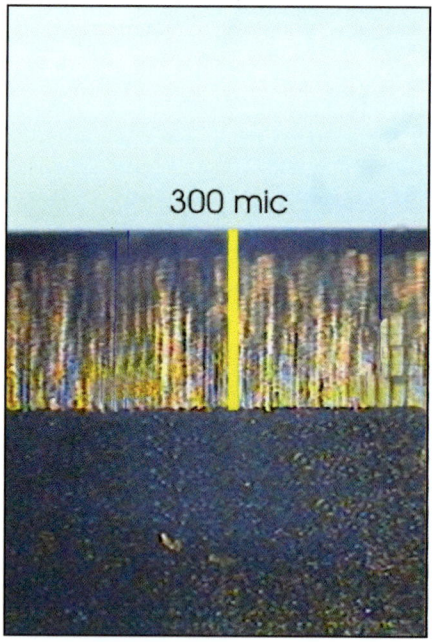

300 mic

Generally the thinnest section is the best. But after several attempts you should be convinced that a medium thick section is perfect for many tasks. If it is cylindrical (not wedged) the surface can be studied at high powers successfully. And it allows a good use of the COL, DF and RhQ filters. Many times these are useless when the section is too thin.

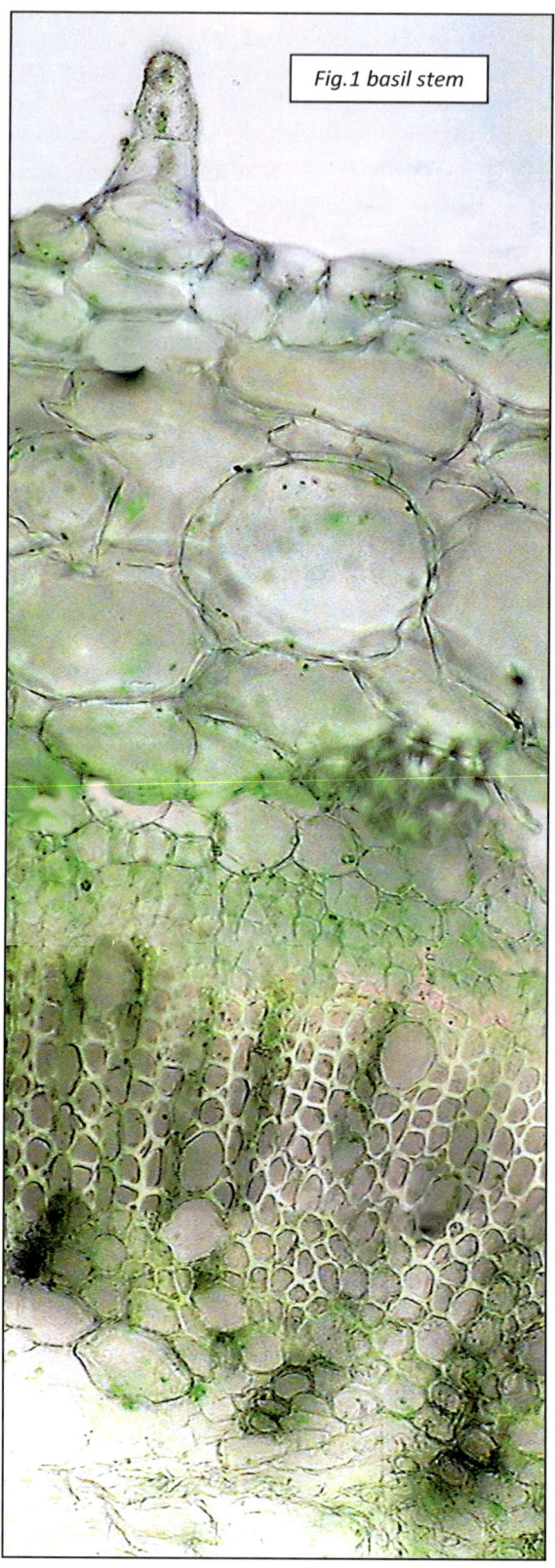

Fig.1 basil stem

But only rarely have plant sections been under 100 microns, and normally they are of 125 to 150 microns. I think this is the result of the sharp edges being V-shaped. Thus the edge is in the centre of the 100 microns blade (50 microns away from any face). If you put in a separator of 50 microns the gap between the sharp edges (not the blades) are 50+50+50 = 150 mic. Theoretically this is the thinnest section you can cut if the blades are parallel (and were not flexible, thus allowing thinner or thicker sections). Do not think of putting the blades together without a separator. Most sections are not hard enough to separate the blades.

I tried several methods to make the internal sides of the sharp edges more or less parallel and closer, but not one of them was successful.

The most easily cut and thinnest sections were obtained using half a razor blade as a separator. And this turned out to be also the easiest way to build the slicer. But some materials, and especially the longitudinal sections of stems, cut better if the separator is the adhesive tape.

So as stated in the first part of this article, now I currently use two versions of the slicer.

The best razor blades. These are the most rigid ones. Try several trade marks if you can. Flexible blades can be separated by the incoming section. This ends in a wedge profile. To study the anatomy of a plant this is not important. But it is for photographic recording. Anyway, even with flexible blades and by making 3 or 4 attempts with different pressures and speeds, should give you a useful section.

Appropriate materials. Highly lignified materials need to be sliced very slowly and with a firm pressure. You may experience great difficulties trying to make sections of Gramineae stems that are very hard. Very soft materials tend to collapse if the blades are too close. The best materials are the medium lignified ones.

Preserving materials. If you are collecting far away from your laboratory or if you want to preserve some materials for future studies, you can fix them. The simplest fixative recommended for botanical materials is 70% ethyl alcohol. Of course some organs, like petals for example can wilt, but their anatomy should be preserved. For anthers, ovaries, root tips, and any reproductive organ Methacarn 95% could be a better choice. After 1 or 2 hours in Methacarn, change to 70% alcohol for one hour and preserve in a second change of 70% alcohol, better with a 1-2% of glycerin. Don't forget to adequately label your sample.

Shown (*previous page Fig.1*) is a section of a basil stem from the exterior epidermis to the central pith parenchyma, captured at x 40 and half reduced, brightfield, no staining,

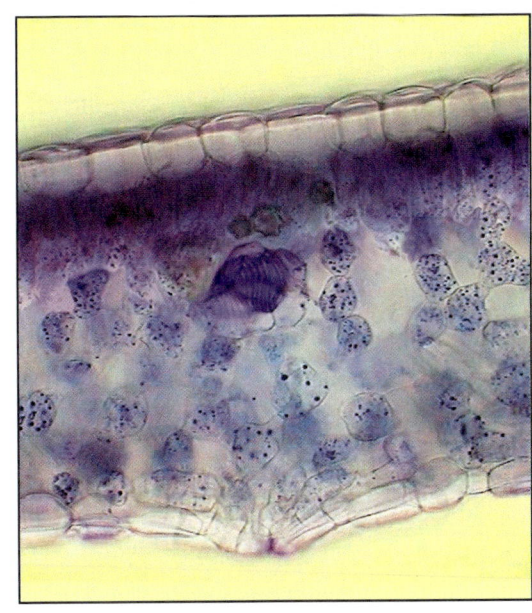

Above: x-section of the lamina of one Aptenia leaf, with a dense palisade parenchyma, and a spongy mesophyl. Hypochlorite - gentian violet.

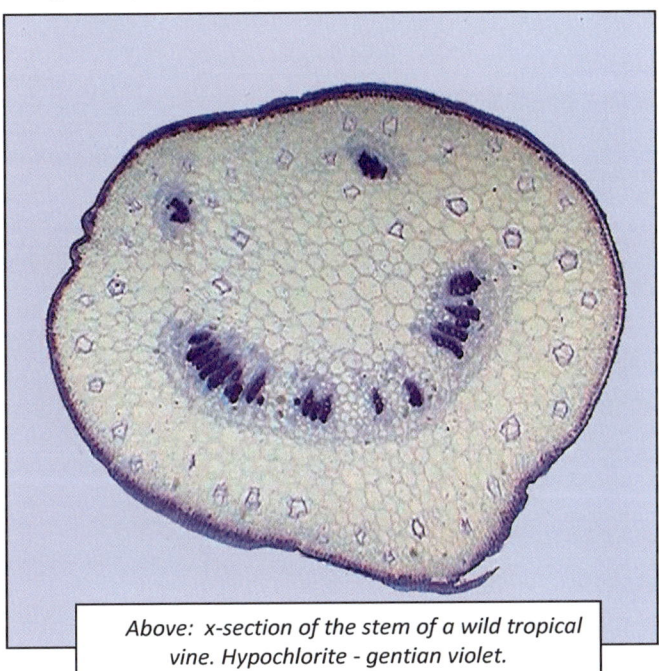

Above: x-section of the stem of a wild tropical vine. Hypochlorite - gentian violet.

Amateur staining.- Professional staining, as you have discovered if you've just read the references, calls for metachromatic dyes like o-toluidine, or for safranin, or double stains such as carmine-green or similar. Or phloroglucynol for lignified tissues. All are difficult to buy or very high priced items.

I have found that a very common and easy to buy dye is gentian violet, a former disinfectant for the babies "mugget", and other infections. Here it is sold as a one percent solution in water. It is very stable; 20 ml is a provision for all your life, because it is used at a drop for every 10 or more millilitres of water. This working solution, which also keeps very well, stains the schlerenchyma and fibers a dark violet, the xylem a more deep color with a red tinge, and the cambium, phloem and the collenchymas a slight purple. The cuticle of the epidermis is also deep colored. As discussed later, also methylene blue can be resorted to.

Contrast Discs.- If you cannot get gentian violet (also

known as crystal violet) you do not need to renounce Technicolor. Your best choice is to take recourse of the contrast discs. The "dispersion staining" that Ted Clarke has appropriately described for DF contrast disks at high powers, is a common characteristic well known to amateurs working at 4x or 10x.

The behavior of the contrast discs is so dependent on the thickness and the nature of the materials that you must play around with your own discs to try the best effect. I have more than 30 different ones. I never know

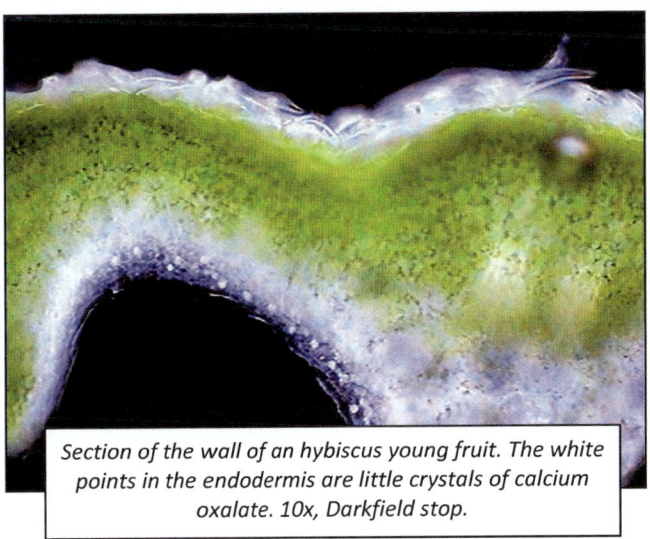

Section of the wall of an hybiscus young fruit. The white points in the endodermis are little crystals of calcium oxalate. 10x, Darkfield stop.

which of these will do the best job with a particular section. But the darkfield stops, some Rheinberg filters and some modifications of the Nomarski simulation filters proposed by Wim van Egmond give outstanding images. The COL filters normally behave only as special darkfield stops.

The better Rheinbergs are blue centered (12-15 mm

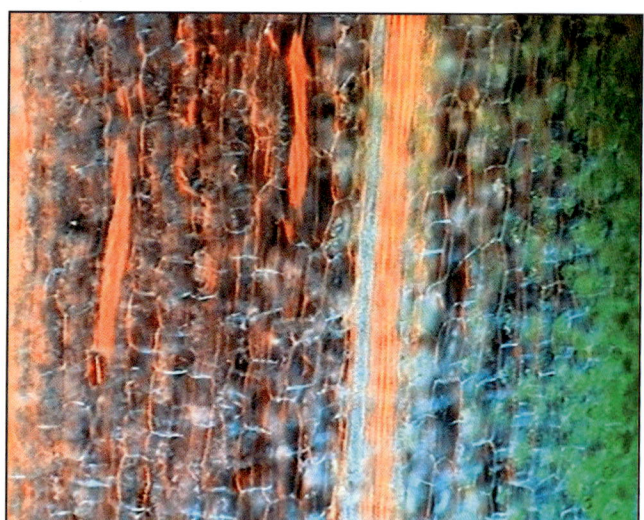

A longitudinal section of the stem of Aptenia. Idioblasts, charged with acicular raphids show a red glow. At right the epidermis. The orange band in the center is the xylem of the vascular bundle. 10x, with a RhQ filter.

diameter) with the exterior ring in different tints of yellow or orange. A very useful modification is a disc with a first clear ring 2 mm wide, a red ring 3 mm wide, a clear one of 2 mm, a blue one of 2 mm and a black centre 12 mm in diameter. Centred or a little displaced it gives many useful nuances. Another remarkable Rheinberg is the Quadrant's Rheinberg (RhQ).

The original van Egmond filters are black discs, with a marginal transparent crescent and a circular blue or purple center. They are a combination of darkfield with Rheinberg and oblique lighting and give its best results with discrete objects like fibres, spicules, sand and the like, especially if they are of high refractive indexes.

I replace the black backgrounds for coloured ones (deep blues and deep reds for example) and make the centers of a diameter similar to the darkfield disk for the objective in use, in a contrasting colour. I leave the transparent crescent unchanged. Its best performance is with relatively thick sections.

These contrast discs (darkfield, Rheinberg and Nomarski simulators) allow the optical differentiation of the different tissues, in medium thin sections, simulating stained sections. They are really very useful (for the 4x and the 10x and with limitations up to the 40x objectives) to give variety and gaiety to the photographed sections. Resolution suffers a little with the 40x. A word of warning: the colours most useful for the visual rendering of sections are aggressive for the sensor of my photographic camera and gives a very bad rendition when compared with the visual image. But they behave very well in direct view. My Col-D3 with a yellow background gives strange results. I see the image in yellow nuances, but the camera records them as many different blue tints. So be prepared, in case your camera behaves in the same way. Judging by published comments most cameras, including the high priced ones,

share this problem.

Anyway the difference between a brilliantly coloured section and a gray and more opaque one is really outstanding.

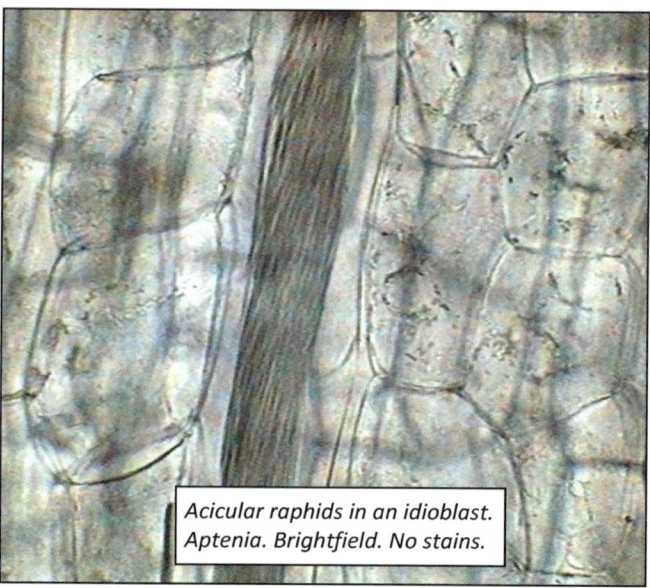

Acicular raphids in an idioblast. Aptenia. Brightfield. No stains.

Calcium oxalate crystals in cortical parenchyma of Aptenia. Brightfield.

Epipremnum x section in brightfield, Iodine added to stain starch.

Uncoloured sections

If you mount your just made sections without any subsequent treatment in glycerin or PVA-G you can make a profound and informative study of the anatomy of an almost living material. Glycerin and PVA-G act as preservatives and even the chlorophyll lasts for many days unchanged. All vegetable tissues are easily recognizable by their morphological traits, and the arrangements of the studied organs are very characteristic. You can discover the idioblasts with its secreted crystals, see even the nucleus of many cells, and the plastids containing oils and starch. If you add to the glycerin a trace of iodine, the starch will be coloured blue and you can easily discover the areas of starch production. All this is lost if you void the cells of its contents with hypochlorite. Note: don't try to add iodine to sections made from materials recently exposed to high levels of sunlight; you risk having a mostly blue and illegible specimen. Try materials exposed to low sun intensities.

Microwave ovens use

Mounting in glycerol, or even in PVA-G can exert on the living cells an excessive osmotic pressure. Delicate materials, like algae, tender hairs, epithelium cells, fungal hyphae and fruiting bodies, and so on, can collapse. You can of course use some fixatives and dehydrating routines for a lengthier mounting in glycerin. But a useful rapid technique is to put the materials or sections, collected in a small Petri dish, or even the recently made slides, on the turntable of the microwave oven and apply a 12 to 20 seconds period of radiation at full power. This enhances the infiltration of the mounting media, evaporates water, gets rid of air bubbles, and the cells become turgid again. Experiment to find the best timing for your own oven. Mine is a 700W model. You can extrapolate suitable times for your oven wattage.

x-section of the leaf of a wild vine treated with a mixed technique: Hypochlorite-gentian violet viewed through a COL-D3 contrast disk. The background was inserted in PhotoPaint.

Article is online at:
www.microscopy-uk.org.uk/mag/artapr04/wdslicerb.html

The Dicotyledon Stem
for the beginner botanist from a beginner botanist

WALTER DIONI   Durango (Dgo) México

Published Feb. 2004
www.microscopy-uk.org.uk/mag/artfeb04/wdstem.html

A usable portion of an otherwise badly cut Portulaca stem. Note the two cell thick epidermis and the continuous ring of xylem.

Portulaca
luz polarizada

Obviously there are many resources devoted to this subject. Many of them are beautifully illustrated with excellent photographs of microscopic preparations made by observing all the "rules of the art": sections made in a professional microtome after paraffin embedding, probably with less than 10 microns of thickness, stained with differential stains for the various tissues, and so on.

If I, a dedicated zoologist, dare to add a new page, it is only to show how a simple tool (a "microtome of a double razor blade") described by Michel Neuberg in the French MICROSCOPIES Magazine, can help an amateur equipped with a microscope and some contrast filters to carry out some investigations and to learn first hand about dicotyledon anatomy... and to support (at almost no cost) *the passing of some days without access to biological water samples.*

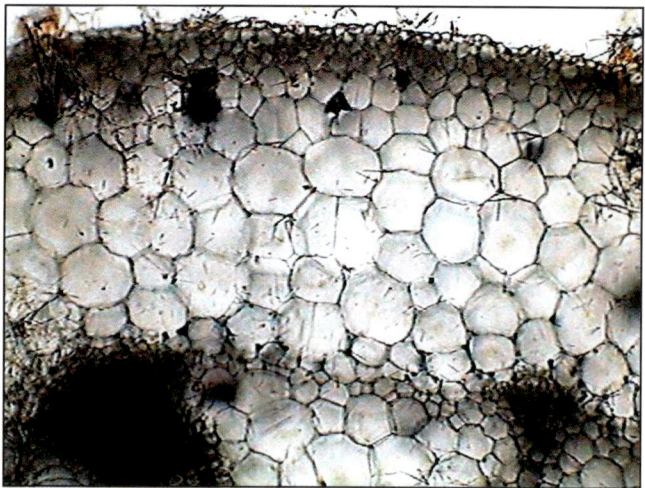

Objective x10: due to thickness of the section, in brightfield **the vascular bundles are opaque.**

It is also true that there are on the Web at least 4 or 5 other designs of differing complexity to build home-made microtomes, or references in almost all the microscopy groups

to the old but useful Ranvier botanist's microtome. But with the exception of Dominic Voisin, a member of the French Microscopies group, there are almost no examples of what the amateur can do with these instruments. I start here by using what is essentially a simple razor blade, but with an ingenious twist.

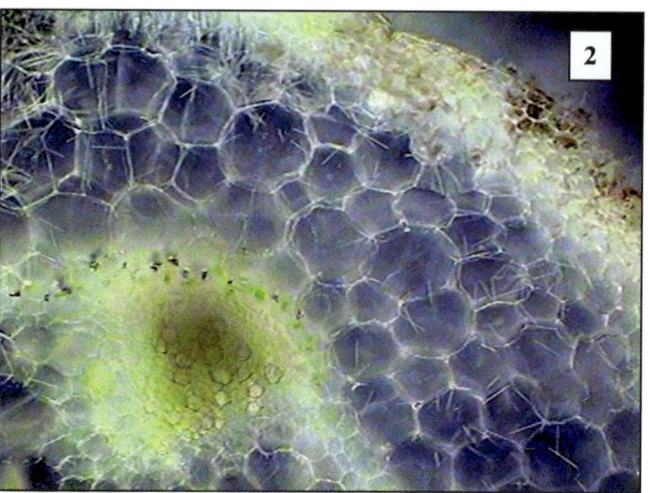

2 - With the COL-D1 filter some tissues are differentiated and the vascular bundle show its structure - X10.

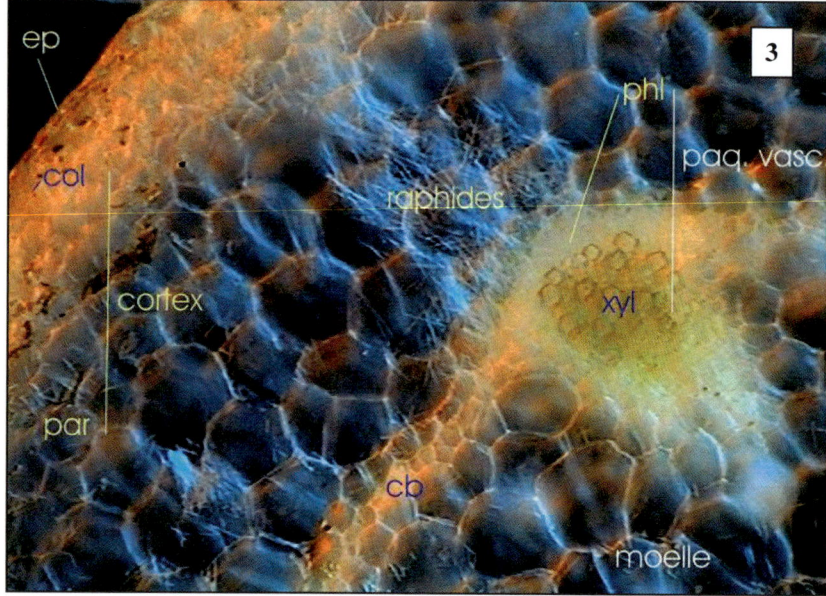

Left.
3 - A similar section labeled for ease of understanding the structure of the stem - X 10 - Filter Rh-Q.

Left.
ep = skin of the stem - col = colenchyma - par = cortical parenchyma (colenchyme + parenchyma form the cortex) -cb = cambium interfasciculaire - Paq.vasc. = package or vascular bundle formed by the phloem (ph) and the xylem (xy). Later it will be seen with more details - moelle, the central cylinder of parenchyma (the pith). Cambium plus pith form a central cylinder: the stele. The raphides are oxalate of calcium needles secreted by some cells. Here, the edges of the razor blades have cut an epithelial cell full of raphides throwing them on the cuted surface (see images 6 to 9.)

The material used is the stem of a small ornamental dicotyledon, soft and flexible. I made transverse sections, and took some additional peels of the stem's epithelium.

The "microtome" was prepared by modifying a little the instructions of Neuberg by covering two double edged razor blades with very thin "Scotch tape" separators. The new blades must be degreased with alcohol before being used. As Neuberg advises I used only the central part of the instrument to obtain thin sections of the stem. The cut must be made by sliding down the blades diagonally. Before cutting, dip the closed instrument in water to wet it. Several perfectly cylindrical cuts, with a neatly cut

surface, were possible, but the cutting edge of the blades disappeared quickly, making further attempts useless.

One must cut over a soft surface only as wide as the "zone of cut", to preserve the edge for further use. The cut must be continued until the section between the two blades is completely released from the two now separated parts of the stem. If this is not done, when opening the microtome, the section will be partially attached to one of the larger stem parts, making it much more difficult to separate without damage.

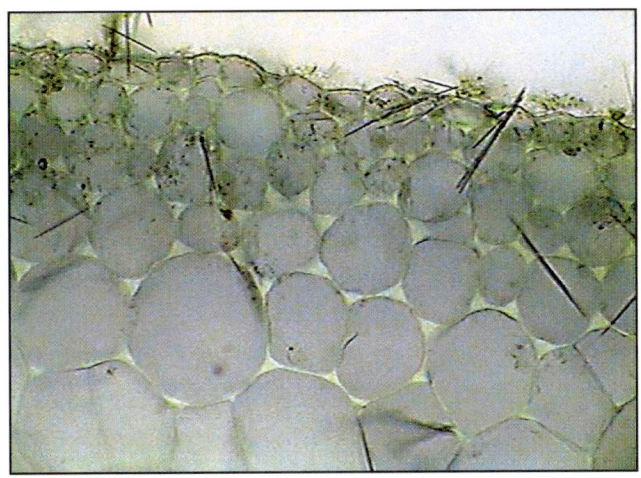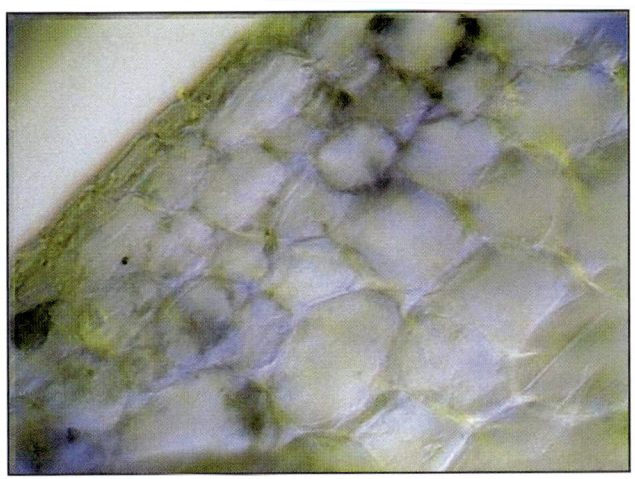

Photographs of the margin of the stem, showing the epithelium in transverse section (forming the outside limit of the stem, with a thickness of 17 to 20 microns), and the subjacent colenchyma. The colenchyma is a support tissue with very thick cellulose walls and specially reinforced corners. Picture at left (fig 4) was taken with the oblique light filter, the one at right (fig 5) with a Rheinberg blue-yellow filter —both with x40 objective.

Any cut carried out with blunt blades will prove deformed and unusable, because the cells would not present a plane face, but one inclined by the action of the defective edge.

I have succeeded to sharpen the blunt blades by passing their edge tilted approx. 30º along the edge of a sharpening stone, three or four times on each face, and then I refined the cutting edge using like a razor strop the palm of my hand. It's up to you to consider if the effort is cost-effective.

If the cut is carried out with a new instrument or one which still preserves a fine edge, even a relatively thick cut can be adequate. With tissues not deformed, the use of all the microscope objectives over the upper face of the cut will give adequate images.

The sections obtained were collected in water, and the thinnest one was chosen and mounted, without any additional treatment, in a 50% glycerin solution, between coverslip and slide. Don't leave the sections to dry at any time or your wet mount will be full of air bubbles. Once the photographic studies were finished, I crushed the cut with two intentions: one, to see the longitudinal aspect of the xylem (see image 20), and another, to be able to measure the thickness of the stem. The flattened stem epithelium showed that the cut had a thickness of around 250 microns!!

Similar cuts and even thinner can be made freehand, but the thicknesses will be more variable, even from one side of the cut to the other, and in my experience it is more dangerous (for the thumb of the potential botanist) than to use the small instrument. The differences in thickness do not matter for the visual study of the materials; this is why the freehand cuts are traditional in the teaching of botany. But for photography they result in a catastrophe. On the Web J.A. Kiernan states that, by trying hard, sections one cell thick can be cut. Using the parenchyma cells of my sample as a rule this implies sections of 150-200 microns. So my cut was not so bad. It was only 1.3 to 1.6 cells deep.

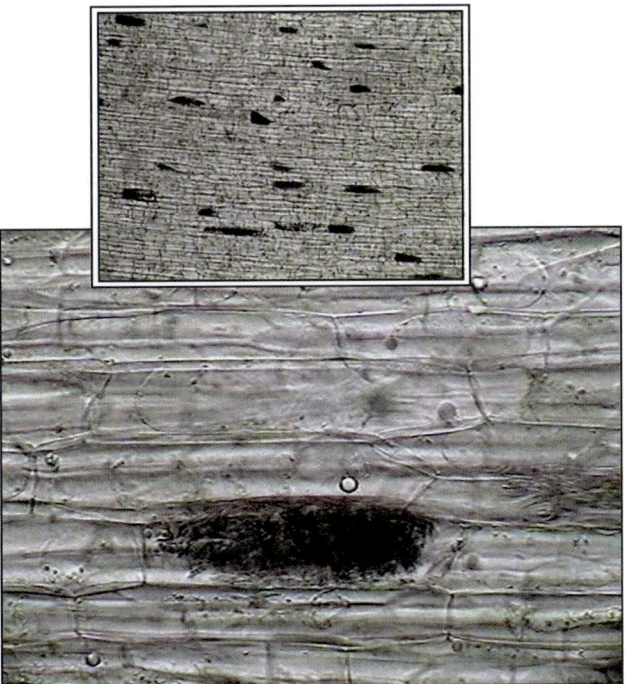

Left.

A piece of the stem skin showing the epithelial cells, X4 (fig. 6) and X40 (fig. 7). Dark spots are specialized cells (called crystal idioblasts) which secrete calcium oxalate needles (raphides). The epithelial cells have a length of 170 to 220 microns and a wide of 25 to 53 microns. Their very refractive nuclei are visible like small circles of 7 to 8 microns

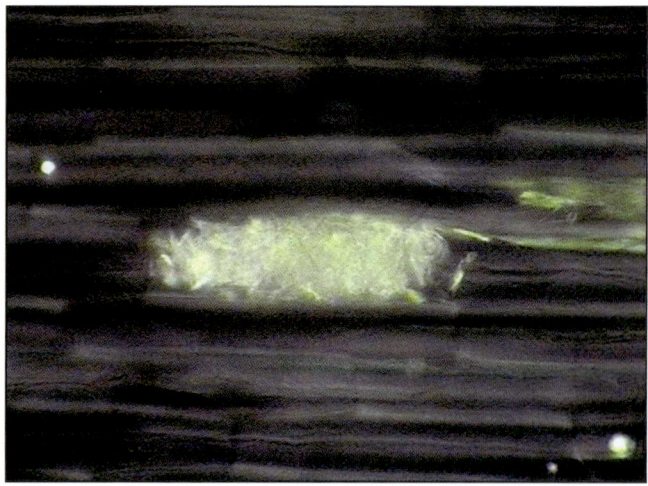 

8 - In polarized light the raphides glow. 9 – On the right a portion of the needles shown in photograph 3, also in polarized light. Both images with the 40x

Theoretically the cuts must be the thickness of the adhesive tapes used as separators. It could be that my razor blades were too flexible, or that I have failed to use an adequate amount of pressure.

Because the blades become blunt, I cannot prove any of these assumptions. One can buy old double edge razor blades at Durango, but it is impossible to do that in very modern Cancún.

With the harder stems of *Portulaca* one can obtain thinner cuts (probably finer than 100 microns) but generally they are incomplete.

The visible anatomical details in my preparations, and the filters and objectives used, are detailed for each image as it is required.

The studied section had a diameter of 4.356 mm. The

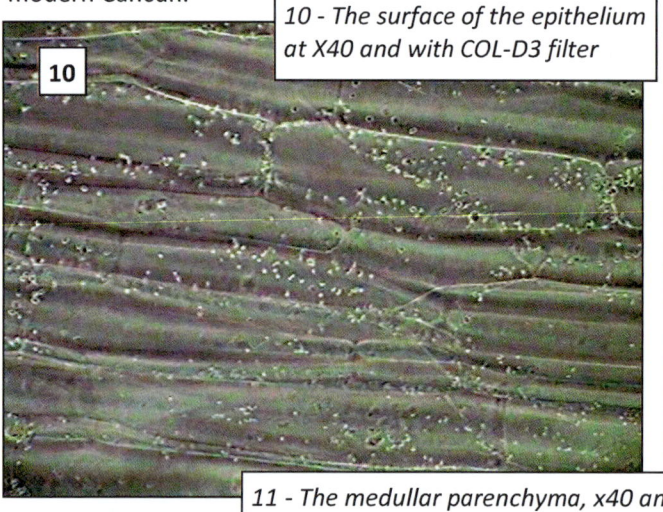

10 - The surface of the epithelium at X40 and with COL-D3 filter

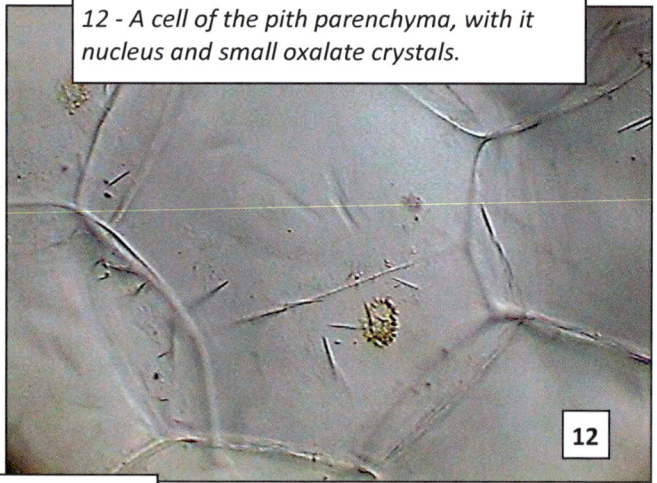

12 - A cell of the pith parenchyma, with it nucleus and small oxalate crystals.

11 - The medullar parenchyma, x40 and with the Rh -Q filter. Small oxalate crystals (the botanist's crystal sand) shine in the cytoplasm.ograph 3, also in polarized light. Both images with the 40x

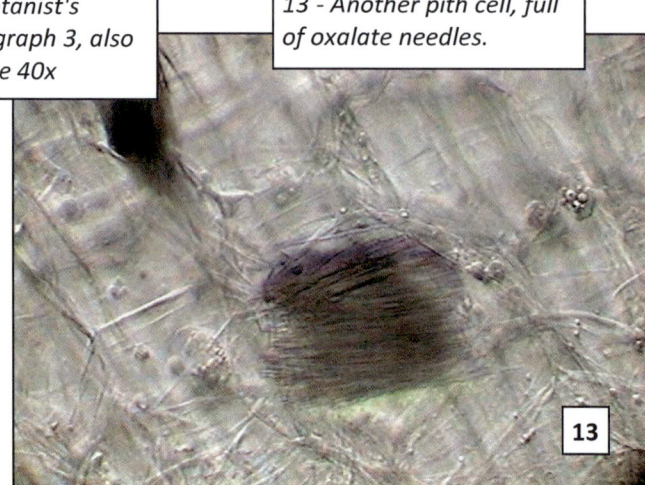

13 - Another pith cell, full of oxalate needles.

stem has an epithelium of only one layer of cells, 17 to 20 microns thick (but 200 microns long, see fig 7), and a cortical area (that between the skin and the interior cylinder of cambium) of 0.535 mm. Cortical cells are ca. 100 microns in diameter. The pith has a uniform parenchyma, has a diameter of 3.700 mm, and its cells have a mean diameter of 180 microns.

All measurements were made from screen images using the software supplied with the microscope.

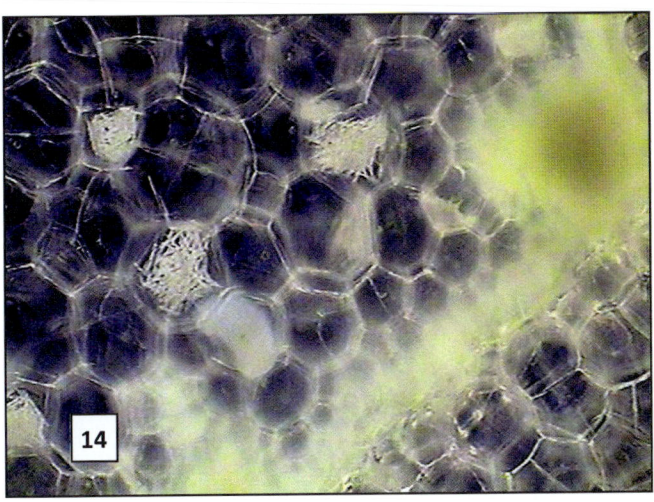

14 – A strip of cambium, separate cortex (below) from pith. Some pith cells show crystal clusters. The image was recorded with the x10 and RhQ filter.

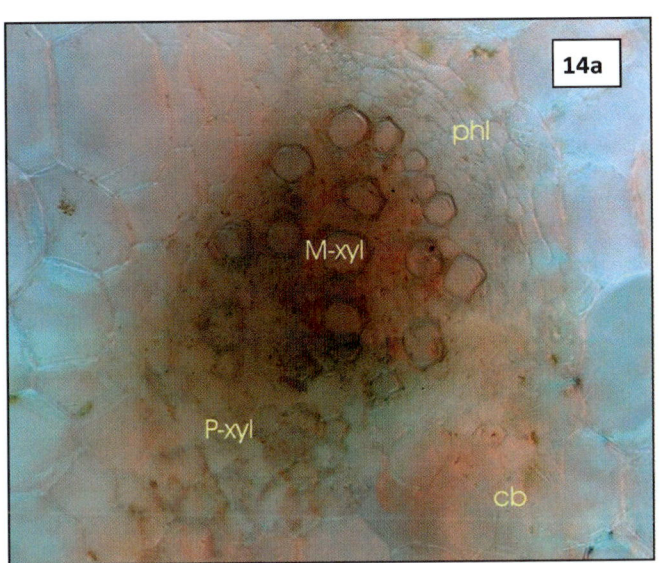

14a - A mosaic of 4 images, to show the structure of the vascular package. Original images were taken with the X40 objective, and COL-D2 filter. The labels are:
phl - phloem,
cb - cambium,
M-xyl - meta xylem,
P-xyl - proto xylem.

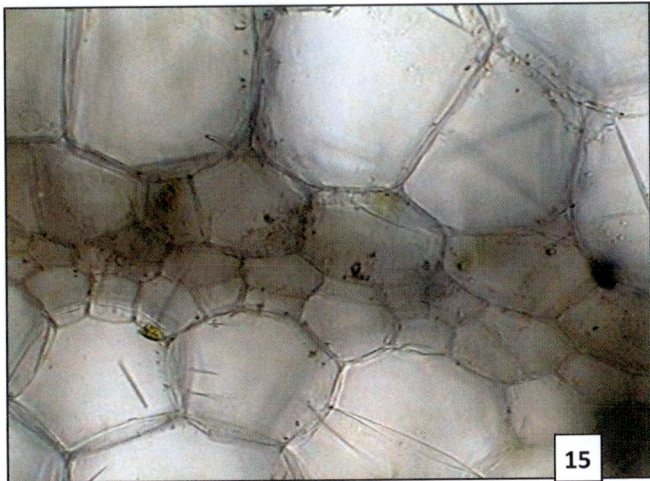

16 - The same sector as in Fig 14, seen in polarized light. As you can see (at least in this material) The RhQ acceptably mimics the polarized light behaviour

15 – The Cambium is a support and generating tissue for the vessels which transport the water minerals and elaborated materials up and down through the stems. It forms a cylinder of 70 to 90 microns thickness between the cortex and pith.

In other species, the strip of cambium supports vessels all over its circumference (see the *Portulaca* section in the title picture). In this species they are concentrated in discrete zones (vascular bundles or packages), which we will see next, and are formed by the phloem, external to the belt of cambium, that conduct solutions downward, and the internal xylem with large vessels that have flexible walls supported by spiral ridges and conduct solutions upward. In other species the phloem is capped by stiff fibres, and it can be a phloem internal to the xylem.

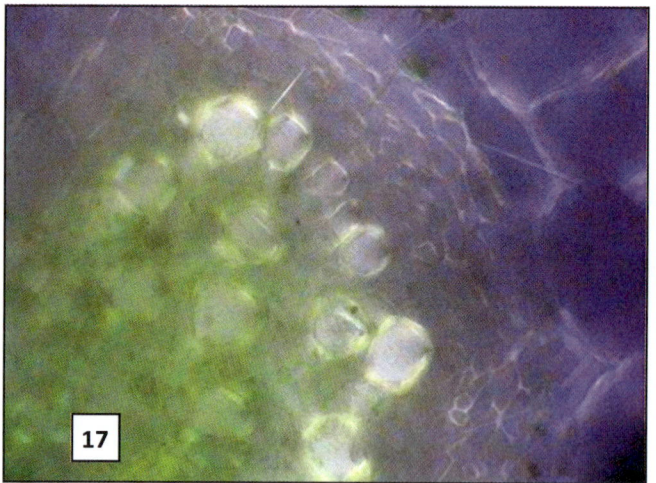

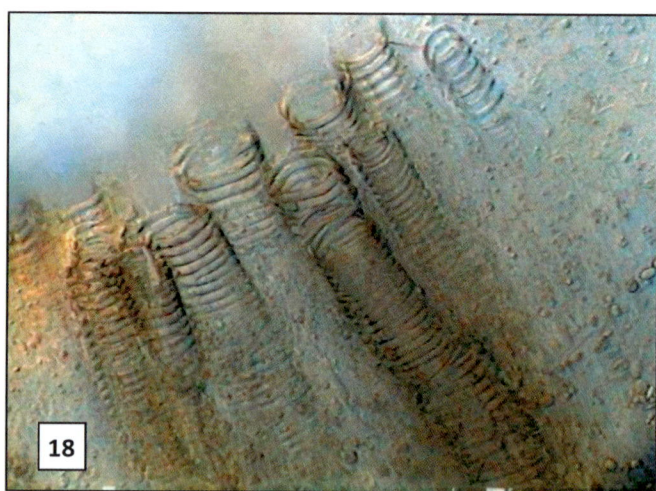

17 - the walls of the vessels are birefringent

*18 - It is reinforced by spiral ridges.
Four levels, amalgamate with Combine Z.*

After this experience I think that especially if Kiernan's statement is true, the use of a single edged razor blade, or (to protect fingers) a better skilled use of the Neuberg double razor blade can open up to the amateur a fascinating research area.

Now, of course, I want to use, one of these days, and to leave testimony of the results, what is generally known as a "nut-and-bolt microtome". It is very simple and a not expensive approximation to the Ranvier botanical microtome.

Who knows? Perhaps I can acquire a second hand old but good *Minot rotary microtome, absolute alcohol, xylene, paraffin and embedding oven, to finish my life as a botanist!*

...The Strange And The Beautiful...
By Ian Walker. UK.
Published March 2004.
www.microscopy-uk.org.uk/mag/artmar04/iwstrange.html

Some of the slides used in this article, the spelling of *Fruxinus excelsius* - third slide from the right is incorrect, it's how I read the mounters writing!

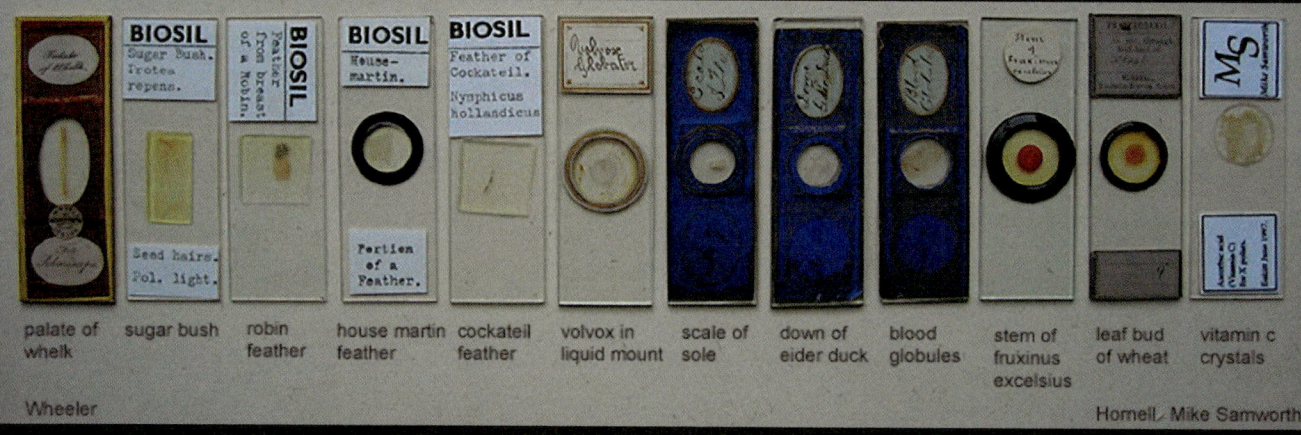

Areas around the cover glass edges, bubbles, drops of oil and poorly preserved specimens can all look interesting under the microscope, especially with dark-field illumination or crossed-polars. Sometimes you can see whole 'scenes' like the four shown after the first image.
....The next time you put that old slide down, have another look you may be surprised at what you might see....

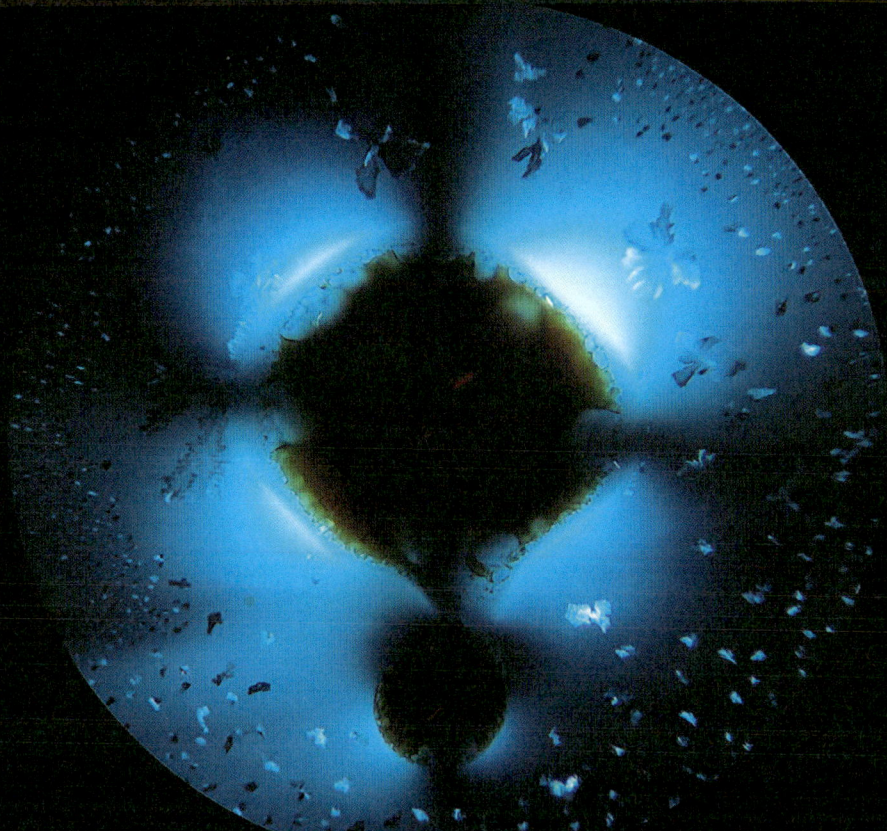

A tiny crystal of Vitamin C under crossed-polars and blue filter, 400x.
Slide prepared by Mike Samworth.

All images taken with the Leica CME microscope and Canon Ixus 400
digital camera using the remote control software.

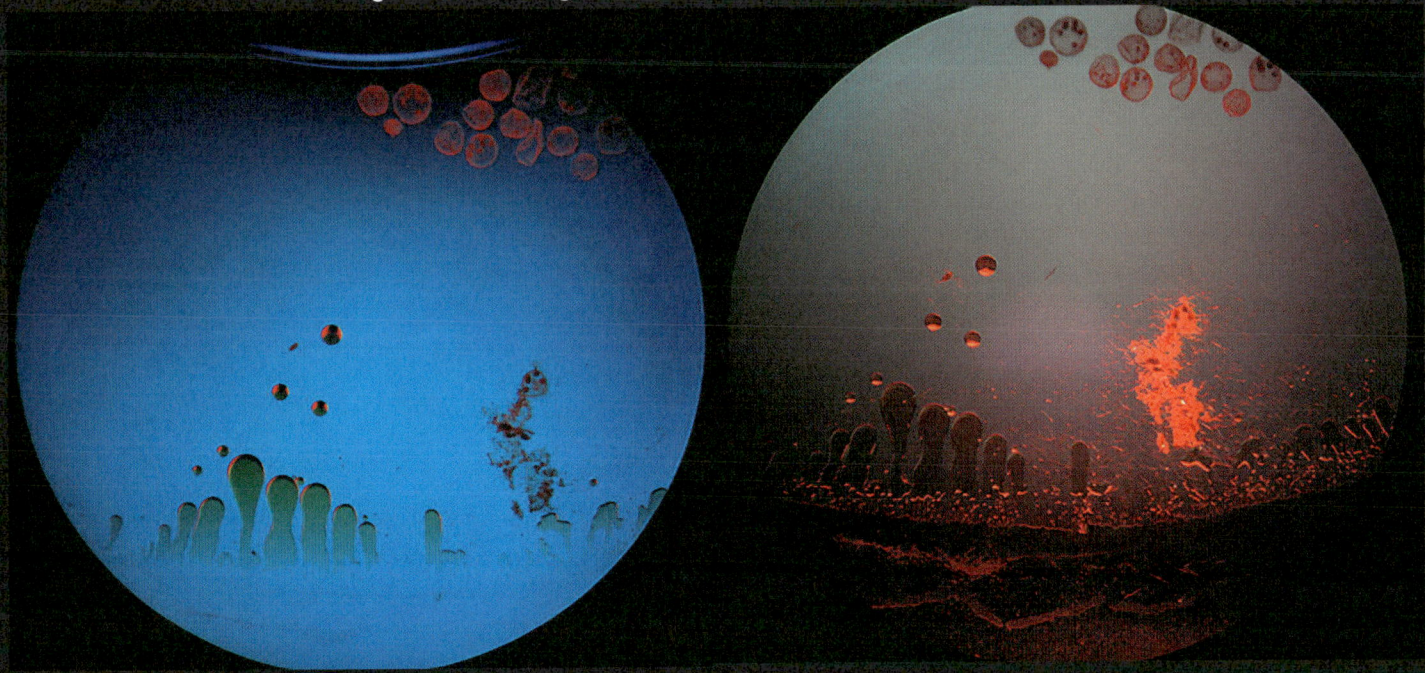

Looking at the very edge of a slide preparation, the tiny creatures at the top are *Volvox* [the main subject matter of the slide] with the left image capturing part of a bubble where the slide has dried out. I am not certain what the 'globules' are but another *Volvox* is entombed in a network of filamentous debris. The two images of the same scene show the dramatic changes that different lighting and contrast techniques can create.   Liquid mount, unknown mounter.

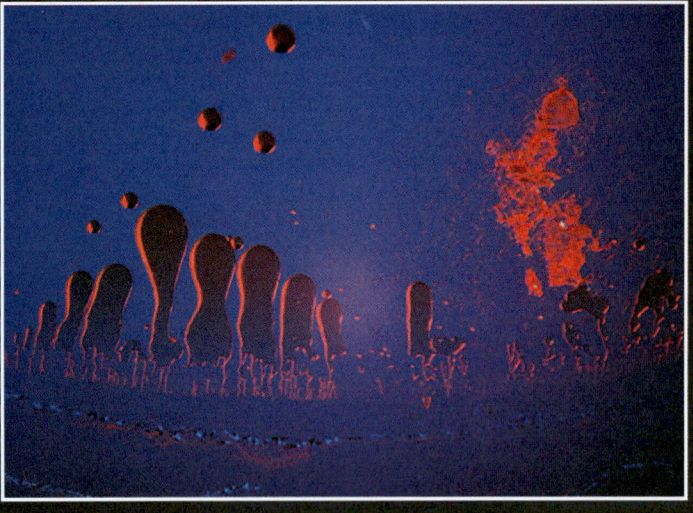
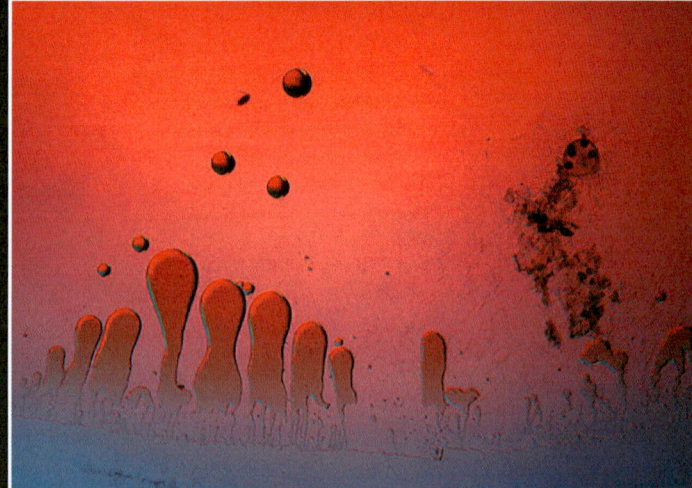

Two more variations using different lighting techniques, this time with the Canon Ixus 400 at a higher zoom level.

The edge of a 'jewel sand' slide under dark-field and Rheinberg illumination.
Slide prepared by J. Percival Yates.

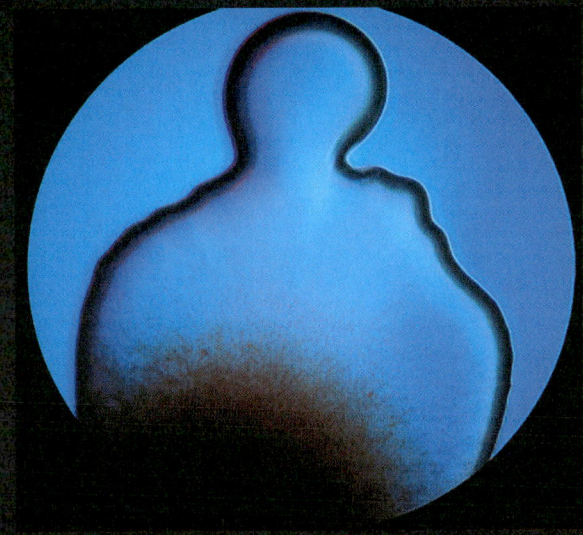

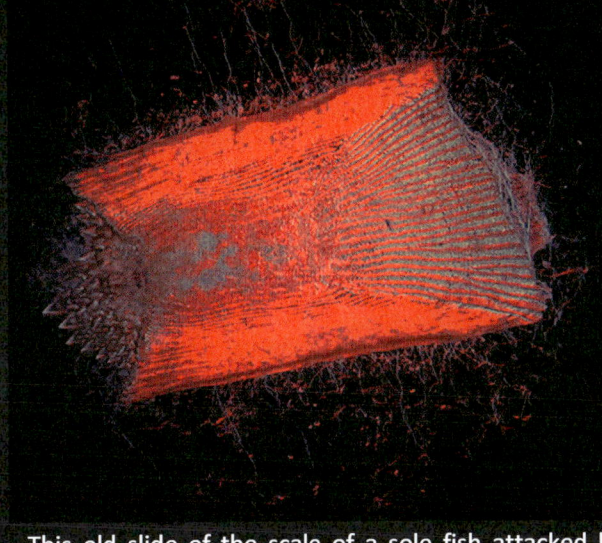

A micro-bubble! 400x.

This old slide of the scale of a sole fish attacked by fungi looks particularly poor under bright-field illumination and could easily be put to the bottom of a slide collection, but it is these very slides where I seek out the unusual. Here the slide comes to life with Rheinberg illumination. I took several of these images using different filters and contrast techniques and each has its own unique character. Old blue paper covered slide by an unknown mounter.

Glycine, 40x.

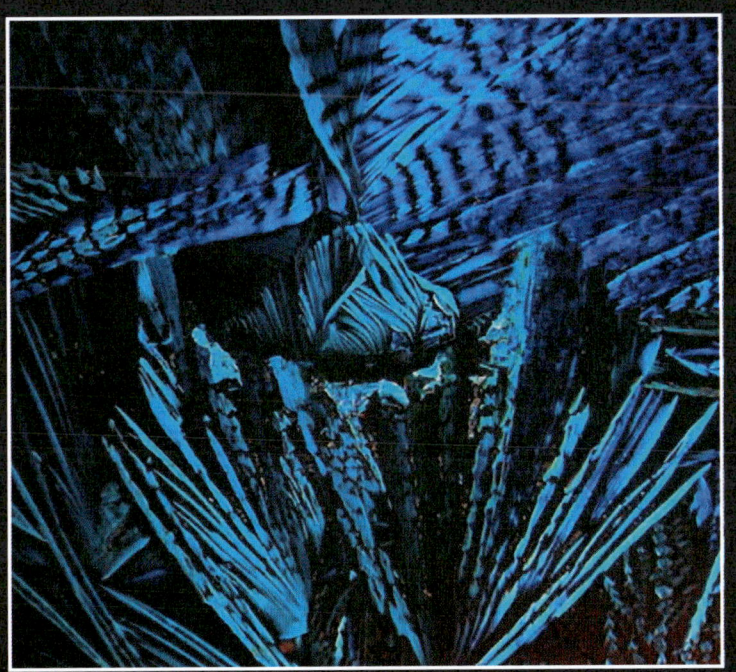

Glycine is the simplest naturally occurring amino acid, it is a sweet tasting crystalline compound found in proteins and is used in the food industry. There are more than 100 amino acids occurring naturally but only about twenty of these are used in building the proteins found in living organisms, we all share these - from the simplest forms of life to plants and animals. Although Glycine is one of the non-essential amino acids required by humans [it is not required from our diet and may be synthesized within the body] it shares the properties of the organic molecules that make up the group and seems fitting that such complex and beautiful crystals seen under the microscope should reflect the diversity of life we see around us. A Biosil slide.

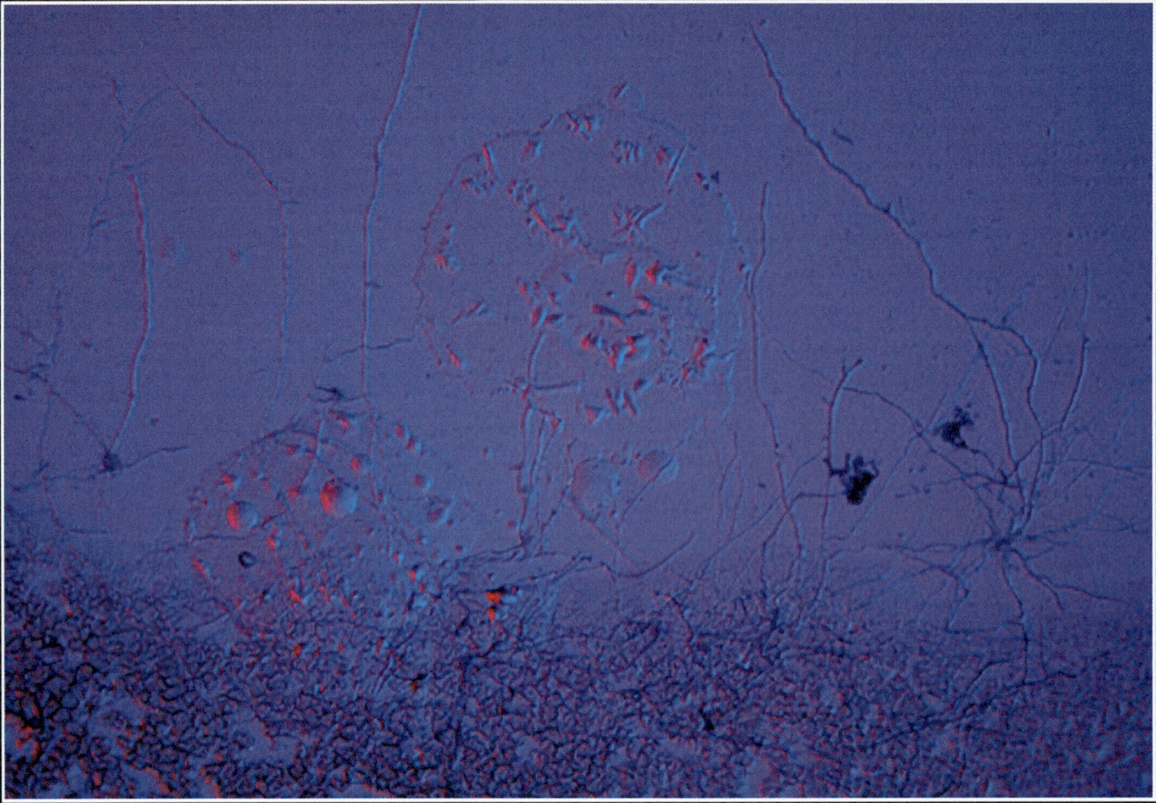

Another fungal attack on the same series of blue paper covered slides as the scale of a sole fish, this time on a slide marked 'Blood Globules', Rheinberg illumination. Old paper covered slide by an unknown mounter.

And now for some more conventional images...

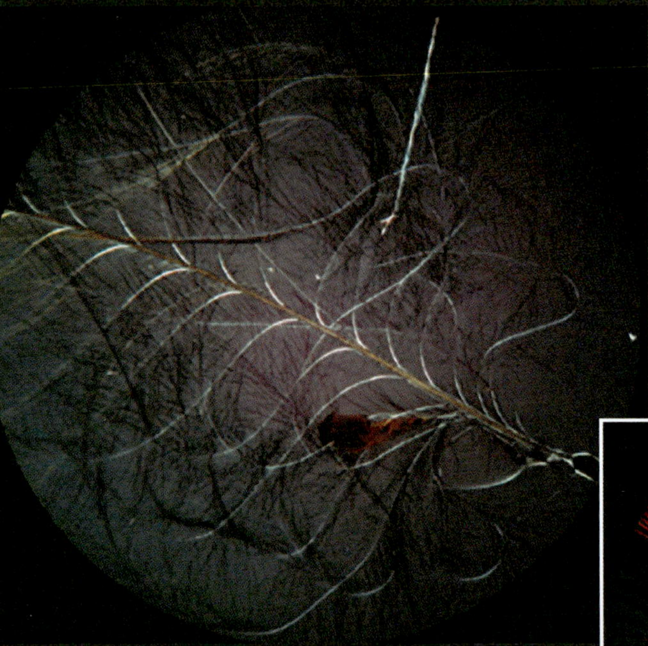

Three feathers. On the left robin, in the centre cockatiel and on the right cardinal bird, all crossed-polars at 40x. Biosil slides.

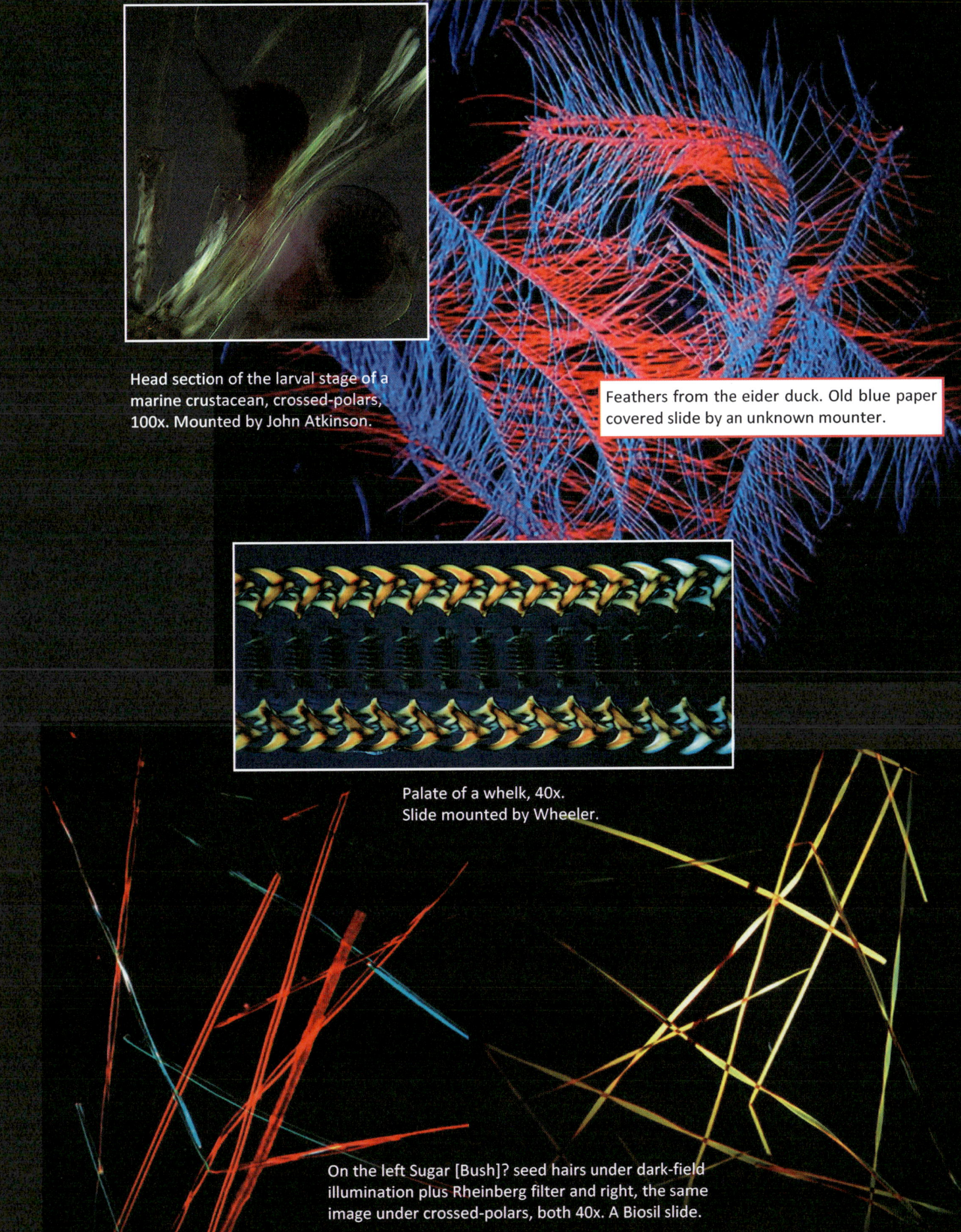

Head section of the larval stage of a marine crustacean, crossed-polars, 100x. Mounted by John Atkinson.

Feathers from the eider duck. Old blue paper covered slide by an unknown mounter.

Palate of a whelk, 40x.
Slide mounted by Wheeler.

On the left Sugar [Bush]? seed hairs under dark-field illumination plus Rheinberg filter and right, the same image under crossed-polars, both 40x. A Biosil slide.

Reflected light image of Polycistina from Barbados, a composite from four images using Combine Z4.2 stacking software by Alan Hadley, 40x. As with a lot of these older opaque slides the subjects are covered with a cover slip, unfortunately these tend to 'fog' and/or crack over many years and together with reflections from the upper surface when using my lamp, reduces the contrast of the image. Unknown mounter.

Reflected light image of *Isthmia nervosa* diatoms, 40x. An excellent but damaged slide mounted by Richard Suter, careful use of Photoshop Elements has brought it back to life!

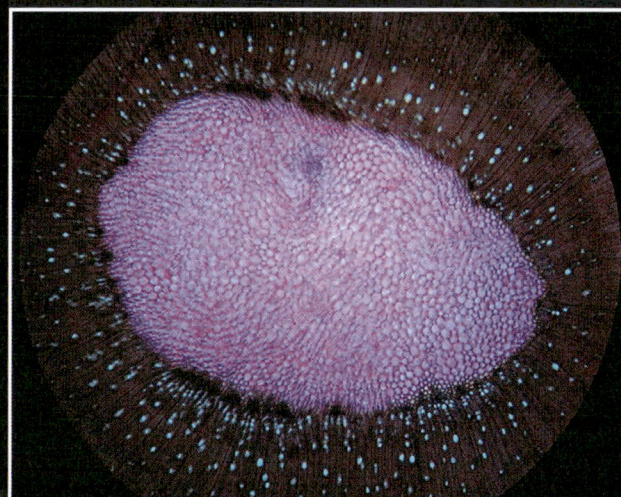

Easter Island also known as Rapa Nui is a remote island in the Pacific Ocean about 2,200 miles west of Chile, formed by a series of underwater volcanic eruptions. It has an area of about 63 square miles and is famous for its massive monolithic stone statues of which some weigh more than 50 tons. Slide prepared by Dr. R. S. Pyne.

Below: The rigid patterns of feathers like this one from a house martin can create a striking effect, upper image at 40x, lower image at 100x. A Biosil slide.

Below: section through the leaf-bud of wheat, dark-field illumination plus Rheinberg filter, 40x. A slide prepared by Hornell, Biological Station Jersey.

Bottom right: dark-field illumination plus blue filter, 40x. Slide by an unknown mounter, cross-section of the stem of *Fraxinus excelsior*, European Ash tree.

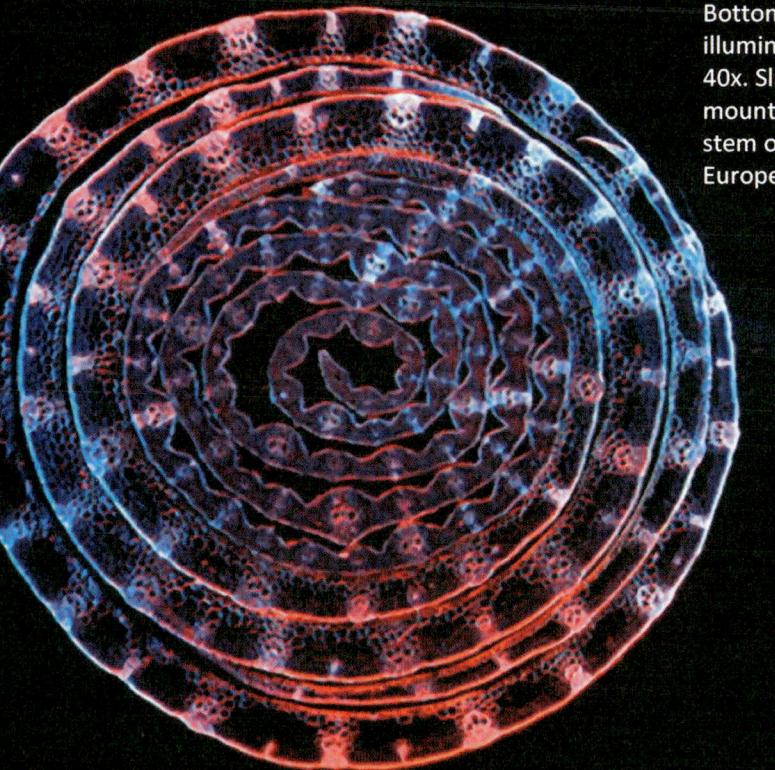

Last Word

Here are then, at the last page of our very first true physical publication. I had sought to include a mix of articles from June 2014 to July 2016, but I found myself also including a few articles outside of that period. It's quite difficult to be just plain logical when you see the wealth of material contributed over the years. The work I included outside of the yearbook's official range was to ensure a small handful of very long-term writers' works were included here... just in case we never get a chance to produce another yearbook!

I suspect many readers would have liked to have seen more 'how to' articles, whilst others would have preferred a greater wealth of imagery, or possibly more historical content. It was left to me alone to try and select the 'right' mix. I do hope I have not disappointed. I would loved to have included everything we ever published in Micscape online, but then I would be sitting here putting it together for the next ten years.

Hopefully, if many people contact us wanting more physical publication of our online content, or take-up of this small work is in sufficient numbers, I can take the time to produce more yearbooks.

As a technical note to anyone interested, the images here were originally published on line as relatively small-sized jpg images at a resolution of 72 DPI. Just lifting them off the web and printing them here at the sizes you witness would just have produced very inferior and heavily pixellated images. That, and the fact many if not most of the images online were never produced or retained at 300 DPI—the minimum resolution required for printing on paper—presented a problem. The solution I used was to 'blow-up' the images using software which employs a fractal-image technique to reproduce the final 300 DPI version of an original 72 DPI image. The software is also used by professional printing houses to achieve the same. The images are unlikely to be of a quality equal to an original large format camera photograph, but hopefully—their quality is not bad when considering they were sourced from far less resolved originals.

Microscopy as a hobby, I think is likely to survive, despite the many distractions in a modern world which steal the attention and focus of a populace towards pursuits offering almost instant gratification. Certainly, the people we have come to know, or were fortunate enough to have brief contact with during the course of running the magazine and web site, have proven to be intelligent, considered, and erstwhile folk. People in future generations will include a minority who share these traits and I trust some of those will come to learn how a microscope and study of the very small can enrich their experience of being a living witness to the extraordinary world we find ourselves in.

In closing, David Walker and myself would like to thank all of our contributors throughout the world for their generous contributions to both the web site, Micscape Magazine and this yearbook. Without you... Without your work collected together in one place... The world of enthusiast microscopy would be far less celebrated or acknowledged.

If you would like to give feedback or comment regarding this yearbook, please use our online contact form at www.microscopy-uk.org.uk or just drop an email directly to me: *molsmith@fastmail.fm*

Mol Smith (on behalf of Mol Smith & David Walker).

Printed in Great Britain
by Amazon